Solving Linear Systems on Vector and Shared Memory Computers

Jack J. Dongarra
University of Tennessee and
Oak Ridge National Laboratory

Iain S. Duff
Rutherford Appleton Laboratory,
CERFACS, and
University of Strathclyde

Danny C. Sorensen
Rice University

Henk A. van der Vorst
Utrecht University

. ®   *Philadelphia*

Society for Industrial and Applied Mathematics

The royalties from the sales of this book are being placed in a fund to help students attend SIAM meetings and other SIAM related activities. This fund is administered by SIAM and qualified individuals are encouraged to write directly to SIAM for guidelines.

Library of Congress Cataloging-in-Publication Data

Solving linear systems on vector and shared memory computers / Jack J.
   Dongarra ... [et al.].
      p.   cm.
   Includes bibliographical references and index.
   ISBN 0-89871-270-X
      1. Algebras, Linear–Data processing. 2. Vector processing
(Computer science) 3. Parallel processing (Electronic computers)
I. Dongarra, J. J.
QA184.S65   1991
512'.5–dc20   90-24045

Contents

Preface

The purpose of this book is to unify and document in one place many of the techniques and much of the current understanding about solving systems of linear equations on vector and shared-memory parallel computers. This book is not a textbook, but it is meant to provide a fast entrance to the world of vector and parallel processing for these linear algebra applications. We intend this book to be used by three groups of readers: graduate students, researchers working in computational science, and numerical analysts. As such, we hope this book can serve both as a reference and as a supplement to a teaching text on aspects of scientific computation.

The book is divided into four sections: (1) introduction to terms and concepts, including an overview of the state of the art for high-performance computers and a discussion of performance evaluation (Chapters 1-4); (2) direct solution of dense matrix problems (Chapter 5); (3) direct solution of sparse matrix problems (Chapter 6); and (4) iterative solution of sparse matrix problems (Chapter 7). Any book that attempts to cover these topics must necessarily be somewhat out of date before it appears, because the area is in a state of flux. We have purposely avoided highly detailed descriptions of popular machines and have tried instead to focus on concepts as much as possible; nevertheless, to make the description more concrete, we do point to specific computers.

Rather than include a floppy disk containing the software described in the book, we have included a pointer to *netlib*. The problem with floppies in books is that they are never around when one needs them, and the software may undergo changes to correct problems or incorporate new ideas. The software included in *netlib* is in the public domain and can be used freely. With *netlib* we hope to have up-to-date software available at all times. A directory in *netlib* called *ddsv* contains the software, and Appendix A of this book discusses what is available and how to make a request from *netlib*.

This book only touches on topics relating to massively parallel SIMD computers and distributed-memory machines, partly because our experience lies in shared-memory architectures and partly because the areas of massively parallel and distributed-memory computing are still rapidly changing. We express appreciation to all those who helped in the preparation of this work, in particular to Gail

Pieper for her tireless efforts in proofreading drafts and improving the quality of the presentation; Ed Anderson, Mary Drake, Jeremy Du Croz, Peter Mayes, Esmond Ng, Al Geist, Giuseppe Radicati, and Charlie Van Loan for their help in proofreading and their many suggestions to improve the readability; and Reed Wade for his assistance in preparing the figures.

Introduction

The recent availability of advanced-architecture computers has had a very significant impact on all spheres of scientific computation including algorithm research and software development in numerical linear algebra. This book discusses some of the major elements of these new computers and indicates some recent developments in sparse and full linear algebra that are designed to exploit these elements.

The two main novel aspects of these advanced computers are the use of vectorization and parallelism, although how these are accommodated varies greatly between architectures. The first commercially available vector machine to have a significant impact on scientific computing was the CRAY-1, the first machine being delivered to Los Alamos in 1976. Thus, the use of vectorization is by now quite mature, and a good understanding of this architectural feature and general guidelines for its exploitation are now well established. However, the first commercially viable parallel machine was the Alliant in 1985, and more massively parallel machines did not appear on the marketplace until 1988. Thus, there remains a relative lack of definition and maturity in this area, although some guidelines on the exploitation of parallelism are beginning to emerge.

We are algebraists rather than computer scientists; as such, one of our intentions in writing this book is to provide the computing infrastructure and necessary definitions to guide the computational scientist and, at the very least, to equip him or her with enough understanding to be able to read computer documentation and appreciate the influence of some of the major aspects of novel computer design. The majority of this basic material is covered in Chapter 1, although we address further aspects related to implementation and performance in Chapters 3 and 4. In such a volatile marketplace it is not sensible to concentrate too heavily on any specific architecture or any particular manufacturer, but we feel it is useful to illustrate our general remarks by reference to some currently existing machines. This we do in Chapter 2 and Appendix C, as well as in Chapter 4 where we present some performance profiles for current machines.

It would be neither practical nor sensible to cover all aspects of parallelism in a book of this size. Instead, we have concentrated on the more well-established area of shared-memory architectures, giving only outline information on distributed-memory and massively parallel architectures.

Linear algebra—in particular, the solution of linear systems of equations—lies at the heart of most calculations in scientific computing. We thus concentrate on this area in this book, examining algorithms and software for dense coefficient matrices in Chapter 5 and for sparse systems in Chapters 6 and 7, where we discuss direct and iterative methods of solution, respectively. Although we have concentrated on this aspect of linear algebra, many of our observations and techniques extend to other areas—for example, the eigenproblem or the solution of least-squares problems, of which brief mention is made in Section 5.5.

Within scientific computation, parallelism can be exploited at several levels. At the highest level a problem may be subdivided even before its discretization into a linear (or nonlinear) system. This technique, typified by domain decomposition, usually results in large parallel tasks ideal for mapping onto a distributed-memory architecture. In keeping with our decision to minimize machine description, we refer only briefly to this form of algorithmic parallelism in the following, concentrating instead on the solution of the discretized subproblems. Even here, more than one level of parallelism can exist—for example, if the discretized problem is sparse. We discuss sparsity exploitation in Chapters 6 and 7.

Our main algorithmic paradigm for exploiting both vectorization and parallelism in the sparse and the full case is the use of block algorithms, particularly in conjunction with highly tuned kernels for effecting matrix-vector and matrix-matrix operations. We discuss the design of these building blocks in Section 5.1 and their use in the solution of dense equations in the rest of Chapter 5. We discuss their use in the solution of sparse systems in Chapter 6, particularly Sections 6.4 and 6.5.

As we said in the Preface, this book is intended to serve as a reference and as a supplementary teaching text for graduate students, researchers working in computational science, and numerical analysts. At the very least, the book should provide background, definitions, and basic techniques so that researchers can understand and exploit the new generation of computers with greater facility and efficiency.

Chapter 1

Vector and Parallel Processing

In this chapter we review some of the basic features of traditional and advanced computers. The review is not intended to be a complete discussion of the architecture of any particular machine or a detailed analysis of computer architectures. Rather, our focus is on certain features that are especially relevant to the implementation of linear algebra algorithms.

1.1 Traditional Computers and Their Limitations

The traditional, or conventional, approach to computer design involves a single instruction stream. Instructions are processed sequentially and result in the movement of data from memory to functional unit and back to memory. Specifically,

- a scalar instruction is fetched and decoded,

- addresses of the data operands to be used are calculated,

- operands are fetched from memory,

- the calculation is performed in the functional unit, and

- the resultant operand is written back to memory.

As demands for faster performance increased, modifications were made to improve the design of computers. It became evident, however, that a number of factors were limiting potential speed: the switching speed of the devices (the time taken for an electronic circuit to react to a signal), packaging

3

and interconnection delays, and compromises in the design to account for realistic tolerances of parameters in the timing of individual components. Even if a dramatic improvement could be made in any of these areas, one factor still limits performance: *the speed of light.* Today's supercomputers have a cycle time on the order of nanoseconds. The CRAY-2, for example, has a cycle time of 4.1 nsec, and Cray Computer Company has announced machines with an expected cycle time of 1 nsec. One nanosecond translates into the time it takes light to move about a foot (in practice, the speed of pulses through the wiring of a computer ranges from 0.3 to 0.9 foot per nanosecond). Faced by this fundamental limitation, computer designers have begun moving in the direction of parallelism.

1.2   Parallelism within a Single Processor

Parallelism is not a new concept. In fact, Hockney and Jesshope point out that Babbage's analytical engine in the 1840s had aspects of parallel processing [94].

1.2.1   Multiple Functional Units

Early computers had three basic components: the main memory, the central processing unit (CPU), and the I/O subsystem. The CPU consisted of a set of registers, the program counter, and one arithmetic and logical unit (ALU), where the operations were performed one function at a time. One of the first approaches to exploiting parallelism involved splitting up the functions of the ALU—for example, into a floating-point addition unit and a floating-point multiplication unit—and having the units operate in parallel.

In order to take advantage of the multiple functional units, the software (e.g., the compiler) had to be able to schedule operations across the multiple functional units to keep the hardware busy. Also, the overhead in starting operations on the multiple units had to be small relative to the time spent performing the operations. Once computer designers had added multiple functional units, they turned to investigating better ways to interconnect the functional units, in order to simplify the flow of data and to speed up the processing of data.

1.2.2   Pipelining

Pipelining is the name given to the segmentation of a functional unit into different parts, each of which is responsible for partial decoding/interpretation and execution of an operation.

The concept of pipelining is similar to that of an assembly line process in an industrial plant. Pipelining is achieved by dividing a task into a sequence of smaller tasks, each of which is executed on a piece of hardware that operates concurrently with the other stages of the pipeline (see Figures

OPERAND 2
OPERAND 1 → STAGE 1) STAGE 2) STAGE 3) STAGE 4) STAGE 5 → RESULTS

Stage 1 = Compare exponents
Stage 2 = Align operands accordingly
Stage 3 = Add exponents and multiply mantissas
Stage 4 = Determine normalization factor
Stage 5 = Normalize result

Figure 1.1: **A simplistic pipeline for floating-point multiplication**

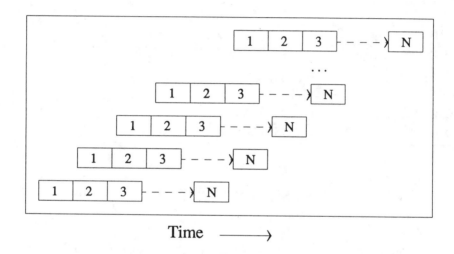

Time ⟶

Figure 1.2: **Pipelined execution of an *N*-step process**

1.1-1.3). Successive tasks are streamed into the pipe and get executed in an overlapped fashion with the other subtasks. Each of the steps is performed during a clock period of the machine. That is, each suboperation is started at the beginning of the cycle and completed at the end of the cycle. The technique is as old as computers, with each generation using ever more sophisticated variations. An excellent survey of pipelining techniques and their history can be found in Kogge [111].

Pipelining was used by a number of machines in the 1960s, including the CDC 6600, the CDC 7600, and the IBM System 360/195. Later, Control Data Corporation introduced the STAR 100 (subsequently the CYBER 200 series), which also used pipelining to gain a speedup in instruction execution. In the execution of instructions on machines today, many of the operations are

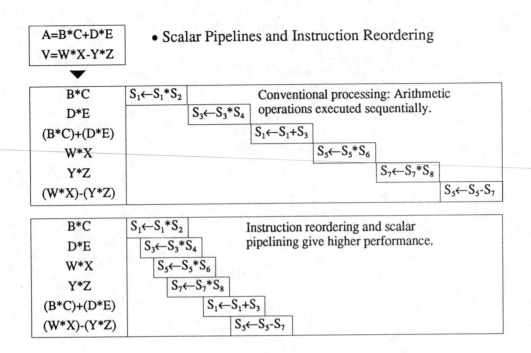

Figure 1.3: **Scalar pipelines**

pipelined—including instruction fetch, decode, operand fetch, execution, and store.

The execution of a pipelined instruction incurs an overhead for filling the pipeline. Once the pipeline is filled, a result appears every clock cycle. The overhead or *startup* for such an operation depends on the number of stages or segments in the pipeline.

1.2.3 Overlapping

Some architectures allow for the *overlap* of operations if the two operations can be executed by independent functional units. *Overlap* is similar but not identical to pipelining. Both employ the idea of subfunction partitioning, but in slightly different contexts. Pipelining occurs when *all* of the following are true:

- Each evaluation of the basic function (for example, floating-point addition and multiplication) is independent of the previous one.

- Each evaluation requires the same sequence of stages.

- The stages are closely related.

- The times to compute different stages are approximately equal.

Overlap, on the other hand, is typically used when *one* of the following occurs:

- There may be some dependencies between evaluations.

- Each evaluation may require a different sequence of stages.

- The stages are relatively distinct in their purpose.

- The time per stage is not necessarily constant but is a function of both the stage and the data passing through it.

1.2.4 RISC

In the late 1970s computer architects turned toward a simpler design in computer systems to gain performance. A RISC (reduced instruction set computer) architecture is one with a very fast clock cycle that can execute instructions at the rate of one per cycle. RISC machines are often associated with pipeline implementations since pipeline techniques are natural for achieving the goal of one instruction executed per machine cycle. The key aspects of a RISC design are as follows:

- single-cycle execution (for most instructions, usually not floating-point instructions),

- simple load/store interface to memory,

- register-based execution,

- simple fixed-format and fixed-length instructions,

- simple addressing modes,

- large register set or register windows, and

- delayed branch instructions.

The design philosophy for RISC is simplicity and efficiency. That is, RISC makes efficient use of the hardware through a simplification of the processor's instruction set and execution of the instruction set.

It can be argued that as super-scalar processors, the RISC processors are close to matching the performance level of vector processors with matched cycle times. Moreover, they exceed the performance of those vector processors on non-vector problems.

1.2.5   VLIW

Very long instruction word (VLIW) architectures are reduced instruction set computers with a large number of parallel, pipelined functional units but only a single thread of control. VLIWs provide a fine-grained parallelism. In VLIWs, every resource is completely and independently controlled by the compiler. There is a single thread of control, a single instruction stream, that initiates each fine-grained operation; any such operations can be initiated each cycle. All communications are completely choreographed by the compiler and are under explicit control of the compiled program. The source, destination, resources, and time of a data transfer are all known by the compiler. There is no sense of packets containing destination addresses or of hardware scheduling of transfers.

Such fine-grained control of a highly parallel machine requires very large instructions, hence the name "very long instruction word" architecture. These machines offer the promise of an immediate speedup for general-purpose scientific computing. But unlike previous machines, VLIW machines are difficult to program in machine language; only a compiler for a high-level language, like Fortran, makes these machines feasible.

1.2.6   Vector Instructions

One of the most obvious concepts for achieving high performance is the use of *vector instructions*. Vector instructions specify a particular operation that is to be carried out on a selected set of operands (called vectors). In this context a vector is an ordered list of scalar values and is inherently one dimensional. When the control unit issues a vector instruction, the first element(s) of the vector(s) is (are) sent to the appropriate pipe by way of a data path. After some number of clock cycles (usually one), the second element(s) of the vector(s) is (are) sent to the same pipeline using the same data path. This process continues until all the operands have been transmitted.

Vector computers rely on several strategies to speed up their execution. One strategy is the inclusion of vector instructions in the instruction set. The issue of a single vector instruction results in the execution of all the component-wise operations that make up the total vector operation. Thus, in addition to the operation to be performed, a vector instruction specifies the starting addresses of the two operand vectors and the result vector and their common length.

The time to execute a vector instruction is given by

```
startup_time + vector_length
```

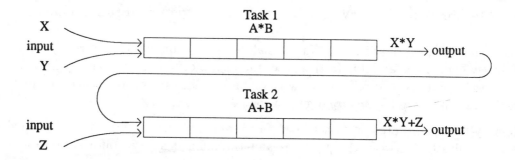

Figure 1.4: **Chaining multiplication and addition**

The time to complete a pipelined operation is a function of the startup time and the length of the vector. The time to execute two overlapped vector operations is given by

`startup_time_2 + vector_length`

The `startup_time_2` is equal to the maximum startup time of the two operations plus one cycle, assuming the independent operation can be initiated immediately after the first has started.

1.2.7 Chaining

Pipelined processes can often be combined to speed up a computation. *Chaining*, or linking, refers to the process of taking the result of one pipelined process and directing it as input into the next pipelined process, without waiting for the entire first operation to complete. The operations that chain together vary from machine to machine; a common implementation is to chain multiplication and addition operations (see Figure 1.4).

If the instructions use separate functional units (such as the addition and multiplication units in Figure 1.4), the hardware will start the second vector operation while the first result from the first operation is just leaving its functional unit. A copy of the result is forwarded directly to the second functional unit, and the first execution of the second vector is started. The net effect is that the execution of both vector operations takes only the second functional unit startup time longer than the first vector operation.

1.2.8   Memory-to-Memory and Register-to-Register Organizations

In some computers, the floating-point functional units may communicate directly with main memory to receive and transfer data. In this case, the operands flow from memory into the functional units and the results flow back to memory as one user-issued operation. This architecture is often referred to as a *memory-to-memory* organization.

Memory-to-memory organization allows source operands and intermediate and final results to be retrieved directly between the pipelines and the main memory. Information on the starting point of the vector, the distance between vector elements (increment), and the vector length must be specified in order to transfer streams of data between the main memory and the pipelined functional units. The CYBER 205 and ETA-10 were examples of this organization.

In contrast to memory-to-memory organization, the floating-point units may have a path to a set of *vector registers,* each register holding several entries. The functional unit then interfaces with the vector registers, and the vector registers in turn have a path to the memory system. The majority of vector machines use this organization, which is often referred to as a *register-to-register* organization.

The presence of vector registers may reduce the traffic flow to and from main memory (if data can be reused), and their fast access times assist in reducing the startup time, by reducing the memory latency of data arriving from main memory. In some sense, then, the vector registers are a fast intermediate memory.

1.2.9   Register Set

Scalar registers are a form of very-high-speed memory used to hold the most heavily referenced data at any point in the program's execution. These registers can send data to the functional units in one clock cycle; this process is typically an order of magnitude faster than the memory speed. As one might expect, memory with this speed is expensive; and as a result, the amount of high-speed memory supplied as registers is limited.

In general, the speed of main memory is insufficient to meet the requirements of the functional units. To ensure that data can be delivered fast enough to the functional units and that the functional units can get rid of their output fast enough, most manufacturers have introduced a special form of very-high-speed memory, the so-called vector register. A *vector register* is a fixed set of memory locations in a special memory under user control.

For example, on Cray computers, a vector register consists of 64 elements. The rule is for a functional unit to accept complete vector registers as operands. The hardware makes it possible for the elements of the vector register to be fed one by one to the functional unit at a rate of exactly one per clock cycle per register. Also, the register accepts 64 successive output elements of the functional unit at a rate of one element per clock cycle. The locations within a vector register cannot be accessed individually.

Loading, storing, and manipulating the contents of a vector register are done under control of special vector instructions. These vector instructions are issued automatically by the compiler, when applicable. Since a functional unit typically involves three vector operands, two for input and one for output, there are more than three vector registers. For example, the Cray processors typically have 8 vector registers. The seemingly extra registers are used by the system in order to keep intermediate results as operands for further instructions. The programmer can help the compiler exploit the contents of these registers by making it possible to combine suitable statements and expressions.

The best source of information about techniques for exploiting the contents of vector and scalar registers is the appropriate computer manual. We also note that the increasing sophistication of compilers is making it less necessary to "help" compilers recognize potentially optimizable constructs in a program.

1.2.10 Stripmining

Vectors too large to fit into vector registers require software fragmentation, or *stripmining*. This is automatically controlled by the compiler. The compiler inserts an outer loop to break the operation into pieces that can be accommodated by the vector register. After the first strip or segment is complete, the next one is started. Because overhead is associated with each vector operation, stripmining incurs a startup overhead for each piece.

1.2.11 Reconfigurable Vector Registers

A special feature of some vector processors is a dynamically reconfigurable vector register set. The length and number of vector registers required usually vary between programs and even within different parts of the same program. To make the best use of the total vector register capacity, the registers may be concatenated into different lengths under software control.

For example, in the Fujitsu VP-200 computer, there are 8192 elements in the vector registers. The registers can be configured into any of the following:

```
# registers x length in words

        32 x 256
        64 x 128
       128 x  64
       256 x  32
       512 x  16
      1024 x   8
```

The length of the vector register is specified in a special register and is set by an instruction generated by the compiler. To best utilize the dynamically reconfigurable vector registers, the compiler must know the frequently used vector lengths for each program. If the vector length is set too short, load-store instructions will have to be issued more frequently. If it is set unnecessarily long, the number of vector registers will decrease, resulting in frequent saves and restores of the vector registers. In general, the compiler puts a higher priority on the number of vectors than on the vector length.

1.2.12   Memory Organization

The flow of data from the memory to the computational units is the most critical part of a computer design. The object is to keep the functional units running at their peak capacity. Through the use of a memory hierarchy system (see Figure 1.5), high performance can be achieved by using *locality of reference* within a program. (By locality of reference we mean that references to data are contained within a small range of addresses and that the program exhibits reuse of data.) In this section we discuss the various levels of memory; in Section 1.4, we give details about memory management.

At the top of the hierarchy are the *registers* of the central processing unit.

The registers in many computers communicate directly with a small, very fast *cache* memory of perhaps several hundred to a thousand words. Cache is a form of storage that is automatically filled and emptied according to a fixed scheme defined by the hardware system.

Main memory is the memory structure most visible to the programmer. Since random access to memory is relatively slow, requiring the passage of several clock cycles between successive memory references, main memory is usually divided into *banks* (see Figure 1.6). In general, the smaller the memory size, the fewer the number of banks.

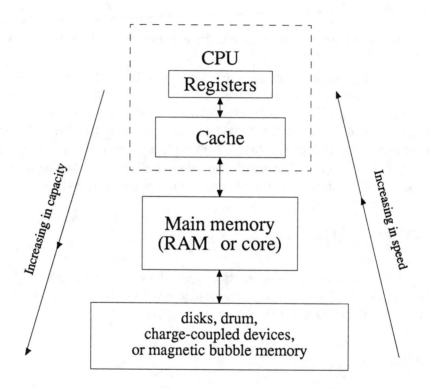

Figure 1.5: **A typical memory hierarchy**

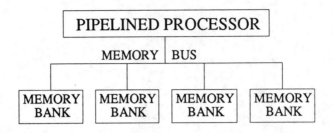

Figure 1.6: **Multiple memory banks in a pipelined computer system**

Associated with memory banks is the memory *bank cycle time*, the number of clock periods a given bank must wait before the next access can be made to data in the bank. After an access and during the memory bank cycle time, references to data in the bank are suspended until the bank cycle time has elapsed. The CRAY Y-MP, for example, has 32 MWords divided into 256 banks, with a bank cycle time of 5 processor cycles (each processor cycle is 6 nsec). This is due to the physical constraint in the number of wires that connect memory to the control portion of the machine. Basically, many locations in memory share common connections.

1.3   Data Organization

Fast memory costs more than slow. Moreover, the larger the memory, the longer it takes to access items. The objective, then, is to match at a reasonable cost the processor speed with the rate of information transfer or the *memory bandwidth* at the highest level.

1.3.1   Main Memory

In most systems the bandwidth from a single memory module is not sufficient to match the processor speed. Increasing the computational power without a corresponding increase in the memory bandwidth of data to and from memory can create a serious bottleneck. One technique used to address this problem is called *interleaving*, or banked memory. With interleaving, several modules can be referenced simultaneously to yield a higher effective rate of access. Specifically, the modules are arranged so that N sequential memory addresses fall in N distinct memory modules. By keeping all N modules busy accessing data, effective bandwidths up to N times that of a single module are possible.

In vector processors, the banks operate with their access cycles out of phase with one another. The reason for such an arrangement is that random access memory is slow relative to the processor, requiring the passage of several clock periods between successive memory references. In order to keep the vector operations streaming at a rate of one word per clock period to feed the pipeline, vectors are stored with consecutive operands in different banks. The phase shift that opens successively referenced banks is equal to one processor clock cycle.

A *bank conflict* occurs if two memory operations are attempted in the same bank within the bank cycle time. For example, if memory has 16 banks and each bank requires 4 cycles before it is ready for another request (a bank cycle time of 4), then for a vector data transfer, a bank conflict will occur if any of 4 consecutive addresses differs by a multiple of 16. In other words, bank conflicts can occur only if the memory address increment is a multiple of 8. If the increment is a multiple of 16, every transferred word involves the same memory bank, so the transfer rate is one-fourth of the maximum transfer rate. If the increment is a multiple of 8 but not a multiple of 16, vector memory transfers occur at one-half the maximum transfer rate, since alternate words use the same bank.

	Banks							
	1	2	3	4	5	6	7	8
L	0	1	2	3	4	5	6	7
o	8	9	10	11	12	13	14	15
c	16	17	18	19	20	21	22	23
a	24	25	26	27	28	29	30	31
t	32	33	34	35	36	37	38	39
i	.	.	.	.	.	.	.	.
o	.	.	.	.	.	.	.	.
n	.	.	.	.	.	.	.	.
s	.	.	.	.	.	.	.	.

Sequential Access with Bank Cycle Time of 4

```
bank 1   0                        8
bank 2     1                        9
bank 3       2                        10
bank 4         3                        11
bank 5           4                        12
bank 6             5                        13
bank 7               6                        14
bank 8                 7                        15
         1  2  3  4  5  6  7  8  9  10 11 12 13 14 15 16 17 18 19
                             clock cycle
```

Every Fourth Element Access with Bank Cycle Time of 4

bank 1	0	8	16	24
bank 2				
bank 3				
bank 4				
bank 5	4	12	20	28
bank 6				
bank 7				
bank 8				

1 2 3 4 5 6 7 8 9 10 11 12 13 14 15 16 17 18 19

clock cycle

Memory bank conflicts cannot occur when processing sequential components of a one-dimensional array or a column of a two-dimensional array in vector mode. However, if equally spaced elements of a one-dimensional array are being processed, bank conflicts are likely to occur if the spacing is a multiple of 8. The effect of such bank conflicts on processing time can be quite pronounced.

A basic attribute used to measure the effectiveness of a memory configuration is the memory bandwidth. This is the maximum number of words that can be accessed per second. The primary factors affecting the bandwidth are the memory module characteristics and the processor architecture.

The processor architecture may also be arranged to ensure a high degree of memory access. Some designs have more than one path to memory. For example, the NEC SX series has multiple paths to fetch one logical vector from memory. In other cases, paths are used for distinct vectors. For example, on the CRAY Y-MP the central memory bandwidth of one stream from memory is one word per cycle of 6 nanoseconds, or 166 million words per second. The Y-MP allows for three streams from memory (two for loads and one for stores) for each processor or 500 million words per second per processor (there can be a maximum of 8 processors on the Y-MP, giving a total of 4 gigawords per second as the peak theoretical memory bandwidth of the system). The memory system is divided into 128 banks and can sustain this rate if there are no bank conflicts.

1.3.2 Cache

Cache memories are high-speed buffers inserted between the processors and main memory to capture those portions of the contents of main memory currently in use. Since cache memories are typically five to ten times faster than main memory, they can reduce the effective memory access time if carefully designed and implemented.

The idea of introducing a high-speed buffer memory between the slow main memory and the arithmetic registers goes back to the ATLAS computer in 1956 [94]. The technique was adopted by IBM for both the System 360 and the System 370 computers. In the IBM System 360 Model 85, for example, the cache (32,768 bytes of 162-nsec semiconductor memory) held the most recently used data in blocks of 64 bytes. If the data required by an instruction was not in the cache, the block containing it was obtained from the slower main memory (4 Mbytes of 756-nsec core storage, divided into 16 different banks) and replaced the least frequently used block in the cache.

Since then, cache memories have been present in most computers from minicomputers to large-scale mainframes. Such a feature is particularly advantageous when memory references concentrate in limited regions of the address space. In such cases, most references will be to data in the cache memory, and the overall memory performance will be effectively that of the faster cache memory.

Cache systems resemble paged systems (see Section 1.4) in basic concept. Their implementation, however, gives rise to great differences in performance. In a paged system, data is retrieved and mapped from disk to main memory by the operating system. In a cache system, data is retrieved and mapped from main memory to a high-speed buffer by the hardware.

Specifically, the process works as follows. When a processor makes a memory request, it generates the address of the desired word and searches the cache for the reference. If the item is found in cache, a *hit* occurs, and a copy is sent to the processor, without a request being made of the main memory (thus taking less time). If the item is not found in cache, a *miss* (or *cache fault*) is generated. The request must then be passed on to the main memory system. When the item is returned to the processor, a copy is stored in the cache, where room must be found for the item. Obviously, cache misses can be very costly in terms of speed.

The speed consideration also dictates that a block of cache, often referred to as a *cache line*, be only a few words long. Cache lines are loaded on demand, and writes (stores or write backs) are usually performed in one of two ways. A block is replaced and written back to memory when a block is about to be discarded or overwritten in cache (referred to as *write back*); or every time a write occurs, the information is written in cache as well as written back in memory (referred to as *write through*).

Cache organizations differ primarily in the way main memory is mapped into a line in cache. In one organization scheme, called *direct mapped cache*, each cache line is mapped into a preassigned location in the cache. For example, for a cache with k blocks, each i, $i + k$, $i + 2 * k$, etc., memory locations will be mapped into block i of the cache. This organization is simple; however, the efficiency of a program may suffer badly from poor placement of data.

In another organization, called *set associative cache*, the cache is partitioned into distinct sets of lines, each set containing a small, fixed number of lines. Each address of main memory is mapped into a particular set. With this scheme the entire cache need not be searched during a reference, only the set to which the address is mapped. When an item that is not in cache is referenced, it is mapped into the first free location in a designated set. If no free location is found, then the least recently used location is overwritten.

1.3.3   Local Memory

On the CRAY-2, there is a "user-managed" cache called *local memory*. Local memory in this case is not a required interface between main memory and registers, as is the cache just described. Rather, it is an auxiliary storage place for vectors and scalars that may be required more than once and cannot be accommodated in the register set. Since the local memory is under user (software) control, it is technically incorrect to call it a cache.

1.4   Memory Management

In many computer systems, large programs often cannot fit into main memory for execution. Even if there is enough main memory for a program, the main memory may be shared between a number of users, causing any one program to occupy only a fraction of the memory, which may not be sufficient for the program to execute. The usual solution is to introduce management schemes that intelligently allocate portions of memory to users as necessary for the efficient running of their programs.

In earlier computers, when the entire program could not all fit into main memory at one time, a technique called *overlays* was used. Portions of the program were brought into memory when needed, overlaying those that were no longer needed. It was the user's responsibility to manage the overlaying of a program.

More common today is the use of *virtual memory*. Virtual memory gives programmers the illusion that there is a very large address space (memory) at their disposal. In the virtual memory concept, the data are stored on *pages*. A page contains a fixed amount of data. For example, in the CYBER 205, data can be stored on "large pages," each of which has a capacity of 65,536 words of 64 bits. The machine instructions refer to virtual addresses of operands, where a virtual address consists of the page number of the page on which the operand is located and the address of the operand on the page. For each program, the operating system keeps a page table which shows where all the pages are physically located (either in main memory or in secondary memory).

When an operand is required that is not available on the pages in main memory, the complete page containing the operand is transported from secondary memory to main memory and thereby overwrites the space of some other page. Usually it overwrites the page that has been least recently used. If data on this page has been changed during its stay in main memory, it is first written back to secondary memory before being overwritten. Transport of a (large) page is called a (large) *page fault.* In organizing an algorithm, one should keep in mind that a page fault takes some I/O time. A good strategy is to carry out as many operations as possible on a block of data, where the block size is chosen such that it fits on the number of pages that one has available during execution.

We illustrate the effect of large page faults with an example given by Winter [172]. This example is for the CYBER 205, but similar effects take place on all virtual memory machines. On the CYBER 205, a large page fault takes about 0.5 second of I/O time. Thus, a large number of large page faults can lead to a large I/O time. Though the virtual memory management is done by the operating system and the user has no information about the pages that are actually in main memory, he can influence the number of large faults.

The example itself concerns the dense matrix-matrix multiplication $C = AB + C$. The matrices are square and of order 1024. Since a large page contains 65,536 elements, it follows that 64 columns of a matrix can be stored on 1 large page and that all three matrices can be stored on 48 large pages in total (about 3 million words). Let us assume that the available main memory space can contain 16 large pages (about 1 million words). This implies that if we access successive columns of a matrix, we have a large page fault after each 64 columns. We now consider three different ways to compute $C = AB + C$.

(a) Innerproduct approach - ijk:

```
      DO 30 I=1,N
        DO 20 J=1,N
          DO 10 K=1,N
            C(I,J)=A(I,K)*B(K,J)+C(I,J)
  10          CONTINUE
  20        CONTINUE
  30  CONTINUE
```

This algorithm is far from optimal for the CYBER 205, for any size of the matrices, since it does not vectorize (the CYBER 205 must have unit increments on the vector in order to vectorize); thus, the CPU time is about 7 minutes. This is a minor problem, however, in comparison with I/O considerations. For each requested row of A, we cause 16 page faults. Since we have to compute about 1 million elements of C, this leads to about 16 million page faults of 0.5 second each. Hence the I/O time is at least 93 days, and in reality the turnaround time will be considerably larger since the machine usually has other work.

(b) Columnwise update of C - jki:

```
      DO 30 J=1,N
        DO 20 K=1,N
          DO 10 I=1,N
            C(I,J)=A(I,K)*B(K,J)+C(I,J)
10          CONTINUE
20        CONTINUE
30  CONTINUE
```

The innermost loop is now a multiple of a vector added to another vector or SAXPY operation, and the two inner loops together represent a matrix-vector product. This approach leads to vector code on the CYBER 205, and hence the CPU time will be in the order of tens of seconds. Again, however, we must consider the I/O time. For each value of J we need access to all the columns of A, which leads to 1024*16 large page faults. Moreover, we need 16 large page faults for B as well as C. The total amount of about 16,000 page faults results in a minimal I/O time of about 2.25 hours. Though this is much better than under (a), an imbalance still exists between CPU and I/O time.

(c) Partitioning the matrices into block-columns - blocked jki:

Now we implicitly partition the matrices in blocks of 256 by 256 each, $nb = 256$. Each block can be placed on precisely 1 large page.

```
  1, N, NB
K = 1, N, NB
 0 JJ = J, J+NB-1, NB
  10 KK = K, K+NB-1, NB
     C(:, JJ) = C(:, JJ) + A(:, KK) * B(KK, JJ)
     NTINUE
     UE
```

spect to the I/O time: the scheme leads to paging A 4 times
and   n only 96 large page faults, and hence takes only a modest 70
secon

Th   ⌄ı designing a multilevel memory hierarchy is to achieve a performance close to that
of the fastest memory, at a cost per bit close to that of the slowest memory.

1.5 Parallelism through Multiple Pipes or Multiple Processors

Machine architects have a number of further options in designing computers with higher computational speeds. These options involve more parallelism in one way or another. The two most common
strategies are to increase the number of functional units (sometimes referred to as *functional pipes*)
in the processor or to increase the number of processors in the system. A third strategy is a hybrid
approach. NEC has a system that features a four-processor machine with 16 arithmetic pipes in
each processor, thus presenting an architecture that utilizes both forms of parallelism.

The multiple-pipe strategy was adopted by CDC in the CYBER 205 and most recently by
the three major Japanese manufacturers Fujitsu, Hitachi, and NEC. In a multiple-pipe machine,
more than one pipe is available for each arithmetic operation. Thus, one can perform an addition
operation, say, on several vectors simultaneously. This may be on independent vectors or on
separate parts of the same vectors where the separation had been performed through stripmining
(Section 1.2.10). In both cases a high degree of sophistication is required in the Fortran compiler,
and there is little (other than the normal vectorization tricks) that the Fortran programmer can
do. This architectural feature is often further complicated by multifunction pipes, where the same
hardware can perform more than one type of arithmetic operation, commonly both an addition
and a multiplication operation.

Though this strategy reduces the CPU time for vector instructions by roughly the number of
pipelines involved, it should not be confused with parallelism at the instruction level. The multiple

pipelines cannot be controlled separately; they all must share the work coming from one single vector instruction. The larger the number of pipelines, the longer the vector length must be in order to get reasonable computing speeds, since the startup time for a single pipeline is more dominant in the multiple situation (see Section 4.4).

Another strategy—the use of multiple processors—was first adopted by Cray Research with the introduction of the X-MP in 1982. Manufacturers have offered products with multiple processors for years; however, until recently these machines have been used principally to increase throughput via multiple job streams. Today, two approaches to multiprocessing are being actively investigated:

1. SIMD (single instruction stream/multiple data stream). In this class, multiple processing elements and parallel memory modules are under the supervision of one control unit. All the processing elements receive the same instruction broadcast from the control unit, but operate on different data sets from distinct data streams. SIMD machines (often referred to as *array processors*) permit explicit expression of parallelism in a program. Program segments that cannot be converted into parallel executable form are sent to the processing units and are executed synchronously on data fetched from parallel memory modules under the control of the control unit. A number of SIMD machines are on the market, including the AMT DAP-610, the Thinking Machines CM-2, and the MasPar MP-1.

2. MIMD (multiple instruction stream/multiple data stream). In MIMD machines, the processors are connected to the memory modules by a network. Effective partitioning and assignment are essential for efficient multiprocessing. Most multiprocessor systems can be classified in this category.

1.6 Interconnection Topology

How information is communicated or passed between processors and memory is the single most important aspect of hardware design from the algorithm writer's point of view. The interconnection network between processors and memories can have various topologies depending on the investment in hardware and desired transfer requirements.

From the physical standpoint, a network consists of a number of switching elements and interconnection links. The two major switching methodologies are *circuit switching* and *packet switching*. In circuit switching, a physical path is actually established between a source and a destination. In packet switching, data is put in a packet and routed through the interconnection network without establishing a physical connection path. In general, circuit switching is much more suitable for bulk data transmission, while packet switching is more efficient for short data messages. A third option is to use both in the same system, producing a hybrid.

The communication links between processing elements fall into two groups, regular and irregular; most are regular. Regular topologies can be divided into two categories:

1. *Static.* In a static topology, links between two processors are passive, and dedicated buses cannot be reconfigured for direct connections to other processors. Topologies in the static category can be classified according to dimensions required for layout, such as one-dimensional, two-dimensional, three-dimensional, and hypercube.

2. *Dynamic.* In a dynamic topology, links can be reconfigured by setting the network's active switching elements. There are three classes: single-stage, multistage, and crossbar.

- A *single-stage* network is composed of a stage of switching elements cascaded to a fixed connection pattern.

- A *multistage* network consists of more than one stage of switching elements and is usually capable of connecting an arbitrary input terminal to an arbitrary output terminal. Multistage networks can be one sided or two sided; a one-sided network has the inputs and outputs on the same side, whereas a two-sided network usually has an input side and an output side and can be divided into three subclasses: blocking, rearrangeable, and nonblocking.

- In a *crossbar* network, every input port can be connected to a free output port without blocking.

1.6.1   Crossbar Switch

The crossbar switch (see Figure 1.7) is the most extensive and expensive connection providing direct paths from processors to memories. With p processors and m memory modules, a crossbar connecting them would require $p * m$ switches, where a switch allows a path to be established allowing data to flow from one part of the machine to another. Contention will occur if two or more accesses are made to the same memory, offering minimal possible contention but with high complexity.

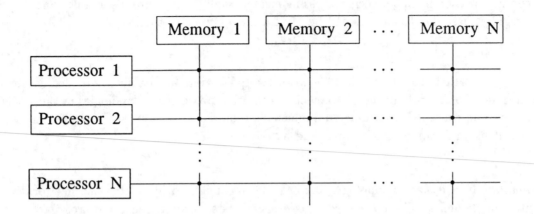

Figure 1.7: **Crossbar switch**

1.6.2   Timeshared Bus

Perhaps the simplest way to construct a multiprocessor is to connect the processors to a shared bus (see Figure 1.8). Each processor has access to the common bus, which is connected to a central memory or memories. This configuration allows for easy expansion.

However, one must be cautious about possible degradation in performance with large numbers of processors. Processor synchronization is achieved by reading from and writing to shared-memory locations. As the number of processors increases, there is a tendency for those shared locations to receive an increasing proportion of the memory references. The performance of the system may degrade rapidly if the data transfer rate on the bus, referred to as the *bus bandwidth*, is not able to deliver data to accommodate the processors. Typically, therefore, bus connections are limited to a modest number (< 30) of processors. Caches and local memory may also be used to help relieve the bandwidth bottleneck.

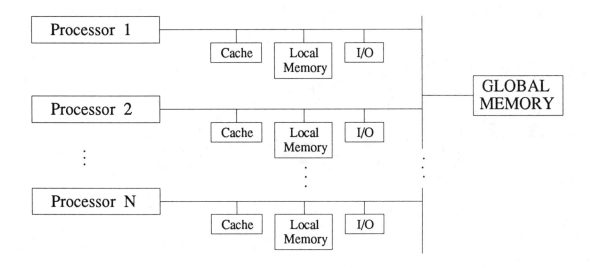

Figure 1.8: **Bus-connected system**

1.6.3 Ring Connection

A ring-connected architecture provides point-to-point connections between processors as well as a cyclic interconnection scheme. Processors place on the ring a message containing the destination address as well as the source address. The message goes from processor to processor until it reaches the destination processor. The advantage of a ring is that the connections are point-to-point and not bus connected. To take advantage of the ring, one can treat the architecture as if it were a pipeline. The effective bandwidth can be utilized as long as computations keep the pipeline filled.

A simple variant is a linear array, where a processor is connected to its two nearest neighbors except at the ends where there is only one connection (see Figure 1.9).

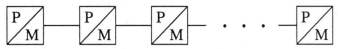

Figure 1.9: **Linear array (P = processor, M = memory)**

1.6.4 Mesh Connection

A simple extension of a linear array of processors is to connect the processors into a two-dimensional grid, where each processor can be thought of as being connected to a neighbor on the north, south, east, and west. The maximum communication length for p processors connected in a square mesh is $O(\sqrt{p})$.

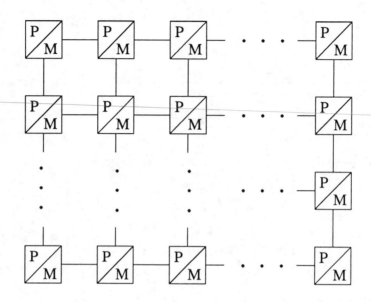

Figure 1.10: **Mesh connection (P = processor, M = memory)**

A variant of this, adopted by machines such as the AMT DAP, is to connect the processors in the first column of Figure 1.10 to those in the last, and those in the first row to those in the last, yielding a toroidal topology.

1.6.5   Hypercube

The hypercube derives its name from the direct connection network used to interconnect its processors or nodes. There are $N = 2^n$ nodes, each of which is connected by fixed communications paths to n other nodes. The value of n is known as the dimension of the hypercube. If the nodes of the hypercube are numbered from 0 to $2^n - 1$, then the connection scheme can be defined by the set of edges that can be drawn between any two nodes whose numberings differ by one bit position in their binary representations. The hypercube design has been used by a number of vendors to form a loosely coupled, distributed-memory, message-passing concurrent MIMD computer. An example of a hypercube topology for $n = 4$ is shown in Figure 1.11.

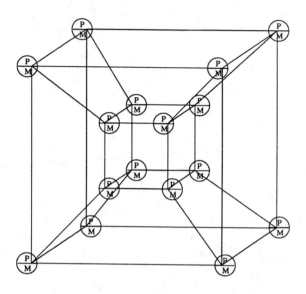

Figure 1.11: **Hypercube connection**

1.6.6 Multistaged Network

A multistaged network is capable of connecting an arbitrary processor to an arbitrary memory. Generally, a multistaged network consists of n stages where $N = 2^n$ is the number of input and output lines. The interconnection patterns from stage to stage determine the network topology. Each stage is connected to the next stage by N paths (see Figure 1.12).

This strategy approximates the connectivity and throughput of a crossbar switch while reducing its cost scaling factor from N^2 to $N \log N$, at the price of an increase in the network latency of $O(\log N)$.

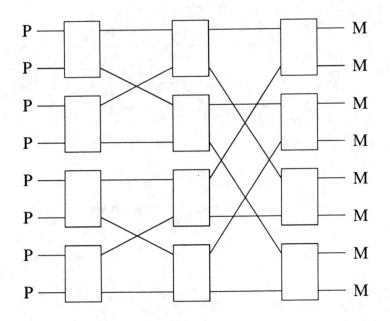

Figure 1.12: **Shuffle-exchange network**

The tradeoffs among some of the different network connections are given in Table 1.1.

Table 1.1: **Tradeoffs among Different Network Connections for p Processors**

Network	Minimum Latency	Maximum Bandwidth per Processor	Wires	Switches	Example
Completely connected	Constant	Constant	$O(p^2)$		
Crossbar	Constant	Constant	$O(p)$	$O(p^2)$	Cray
Bus	Constant	$O(1/p)$	$O(p)$	$O(p)$	Sequent
Mesh	$O(\sqrt{p})$	Constant	$O(p)$		AMT DAP
Hypercube	$O(\log p)$	Constant	$O(p \log p)$		Intel iPSC
Switched	$O(\log p)$	Constant	$O(p \log p)$	$O(p \log p)$	BBN TC2000

1.7 Programming Techniques

Much of the discussion in Chapters 5 through 7 concerns the efficient programming of vector and parallel computers. Here we concentrate on a few basic ideas.

The first observation is as much algorithmic as programming oriented: the programmer should take advantage of any independence within the calculation by isolating independent calculations, usually in separate subroutines, thus expediting the use of macroprocessing on the target machine(s). Such a modular approach is now generally recognized as good programming practice even in the serial regime.

At the finer level, parallelism (as far as the Fortran programmer is concerned) is usually handled by the compiler (we include any manufacturer-supplied pre- or post-compiler). The normal source of exploitable parallelism is found within the Fortran DO-loop. Thus the primary concern of the programmer should be to remove barriers to vectorization or parallelism from the DO-loops. We now discuss some of these barriers and ways of overcoming them.

Usually the simpler the DO-loop, the easier it is for the compiler to recognize vectorization or parallelism opportunities. In particular, it is important to explicitly remove from DO-loops calculations that need not be present within them (i.e., loop-invariant constructions). Furthermore, since a relatively small proportion of the code may prevent a DO-loop from vectorizing, it may be advantageous to split the loop into two loops—one vectorizable, the other not. IF statements and indirect addressing, in particular, can be vectorized only in simple cases; often, by splitting the loop, the part inhibiting vectorization in the complicated original loop can be vectorized in its simpler form, so that both parts of the split loop vectorize. For a comparison of various vectorizing compilers, see [20].

On some machines, difficult constructs can be replaced by calls to machine-dependent routines. Examples include the replacement of IF statements by conditional vector merges on Crays or the CYBER 205, or the explicit call to gather/scatter routines for indirect addressing on early versions of Cray supercomputers. We caution, however, that while these replacements may be necessary for optimal implementation on particular architectures, they do reduce portability.

Some manufacturers allow programmer assistance to the compiler through directives inserted as special Fortran comments in the code immediately prior to the relevant DO-loop. Each manufacturer provides its own set of compiler directives; but, since the directives appear as comments in the program, they do not limit the portability of the program. For example, to ensure that the compiler will vectorize the following loop, we inserted the following directives to make the compiler ignore potential recursion within the loop. These directives are for the Alliant, Cray, and NEC compiler, respectively.

```
CVD$ NODEPCHK
CDIR$ IVDEP
*VDIR NODEP
      DO 10 J = K+1, N
         A(I,J) = A(I,J) - A(I,K)*A(K,J)
   10 CONTINUE
```

Although the necessity for these directives decreases as compilers improve, there are cases when they are essential, and we do not discourage their use in general because portability is not compromised. Nevertheless, a much more satisfactory situation would be for the directives to be standardized. See [107] for a comparison of different parallel Fortran dialects.) We note that the use of the standard and standardized building blocks of the Basic Linear Algebra Subprograms, BLAS (discussed at length in Chapter 5), is satisfactory and should not be considered a barrier to portability.

Many programs have nested loops, commonly only to a depth of two or three. Often, it is mathematically immaterial how the loops are ordered, but the performance can be dramatically affected by this order. Since most vectorizing compilers vectorize only on the innermost loop (the IBM VS Fortran compiler is a notable exception), it is common practice to make the innermost loop that of longest vector length. Similarly, most parallelizing compilers parallelize over the outermost loop so that the size of each task or granularity of separate processes is kept high. On the Alliant the most common way of treating nested loops is termed COVI (concurrent outer – vector inner), although other ways of exploiting loop structure are possible. Again, we acknowledge that as compilers become increasingly sophisticated, such *loop inversion* may be done automatically when beneficial.

Another technique often used with nested loops is *loop unrolling*. As with loop inversion, loop unrolling may be done automatically by the compiler. For loop unrolling, the same vector entry is updated in several successive passes of the innermost loop. For example, in the code

```
      DO 200 J=1,N
         DO 100 I=1,M
            V(I) = V(I) + F(I,J)
100      CONTINUE
200 CONTINUE
```

the entries $V(I)$ might typically be fetched and saved at every iteration of the J loop (loop 200). If, however, the J loop is unrolled (here to a depth of 4), the resulting code

```
      DO 200 J=1,N,4
         DO 100 I=1,M
            V(I) = V(I) + F(I,J)
     *                  + F(I,J+1)
     *                  + F(I,J+2)
     *                  + F(I,J+3)
100      CONTINUE
200 CONTINUE
```

provides four times the number of adds for each vector load/store of V and so should perform better for machines with vector registers or a memory hierarchy. The two problems here are that the code can become complicated (sometimes called pornographic) and can confuse rather than help the compiler. Indeed, as compilers improve and as explicit higher-level BLAS are used in algorithm design, loop unrolling is rapidly becoming an obsolete technique [37, 34].

Finally, there are some constructs that no amount of directives can help—for example, forward recurrences. Here only algorithmic changes can help, which usually increase vectorization or parallelism but at the cost of more arithmetic operations. This topic is discussed further in Section 3.4.

A good reference or guidebook to these techniques can be found in [115].

Chapter 2

Overview of Current High-Performance Computers

A much-referenced taxonomy of computer architectures was given by Flynn [72]. He divided machines into four categories: SISD (single instruction stream/single data stream), SIMD (single instruction stream/multiple data stream), MISD (multiple instruction stream/single data stream), and MIMD (multiple instruction stream/multiple data stream). Although these categories give a helpful coarse division (and we, in fact, use these categories throughout our book), the current situation is more complicated, with some architectures exhibiting aspects of more than one category. Indeed, many of today's machines are really a hybrid design. For example, the CRAY X-MP has up to four processors (MIMD), but each processor uses pipelining (SIMD) for vectorization. Moreover, where there are multiple processors, the memory can be local or global or a combination of these. There may or may not be caches and virtual memory systems, and the interconnections can be by crossbar switches, multiple bus-connected systems, timeshared bus systems, etc.

In this chapter we briefly discuss advanced computers in terms of several different categories, more closely related to size and cost than to design: supercomputers, mini-supercomputers, vector mainframes, and novel parallel processors.

2.1 Supercomputers

Supercomputers are by definition the fastest and most powerful general-purpose scientific computing systems available at any given time. They offer speed and capacity significantly greater than the most widely available machines built primarily for commercial use. The term *supercomputer* became

Table 2.1: **Performance Trends in Scientific Supercomputing**

Year	Machine	Speed
1964	CDC 6600	1 Mflops
1975	CDC 7600	4 Mflops
1979	CRAY-1	160 Mflops
1983	CYBER 205	400 Mflops
1986	CRAY-2	2 Gflops
1990-1995		200-1000 Gflops

prevalent in the early 1960s, with the development of the CDC 6600. That machine, first marketed in 1964, boasted a performance of 1 Mflops (millions of floating-point operations per second). (Throughout this book we shall refer to a floating-point operation as either a floating-point addition or multiplication in full precision (usually 64-bit arithmetic).)

During the next fifteen years, the peak performance of supercomputers grew at an extremely rapid rate; and since 1980, that trend has accelerated. Machines projected for 1995 are expected to have a maximum speed of 200 Gigaflops (billions of floating-point operations per second), more than 200,000 times that of the CDC 6600 (see Table 2.1).

By far the most significant contributor to supercomputing in the United States has been private industry. Companies led by Cray Research and Control Data Corporation have devoted their resources to producing state-of-the-art machines that enable scientists and engineers to tackle problems never before possible. It is from these commercial ventures that we have seen the development of computers capable of solving complex numerical and non-numerical problems. Table 2.2 lists the principal supercomputers that have been marketed in the past several years. We include the CRAY-1 largely as a benchmark, since it could not now be considered a supercomputer in terms of performance. The next generation, with higher speed and more parallelism, is already under development.

Table 2.2: **Supercomputers: Past, Present, and Future**

Machine	Maximum Rate, in Mflops	Memory, in Mbytes	Number of Processors
CRAY-1 †	160	32	1
CRAY X-MP †	941	512	4
CRAY-2	1951	4096	4
CRAY Y-MP	2667	256	8
CRAY-3 ‡	16000	16384	16
CRAY C-90 ‡	16000		16
CYBER 205 †	400(a)	128	1
ETA-10P †	333	2048(b)	2
ETA-10Q †	1680	2048(b)	8
ETA-10E †	3048	2048(b)	8
Fujitsu VP-30E *	133	1024	1
Fujitsu VP-50E *	286	1024	1
Fujitsu VP-100E *	429	1024	1
Fujitsu VP-200E *	857	1024	1
Fujitsu VP-400E *	1714	1024	1
Fujitsu VP-2600/20 ‡*	8000	2048	1
Hitachi S-810/20	857	512(c)	1
Hitachi S-820/80	3000	512(c)	1
NEC SX/1E	324	1024(d)	1
NEC SX/1A	650	1024(d)	1
NEC SX/2A	1300	1024(d)	1
NEC SX/3 model 44 ‡	22000	16384	4

(a) 800 Mflops for 32-bit arithmetic
(b) Also 4 Mwords of local memory with each processor
(c) Also a 12-Gbyte extended memory
(d) Also an 8-Gbyte extended memory
†Computer no longer manufactured
‡Computer not yet manufactured
*Marketed in the West by Siemens (the VP-50 to 400 and S series range)

The price of the systems in Table 2.2 depends on the configuration, with most manufacturers offering systems in the $5 million to $20 million range. All use ECL logic with LSI (Large Scale Integration), except the CRAY X-MP, the CRAY-1 built with SSI (Small Scale Integration), and the ETA-10 in CMOS ALSI (Advanced Large Scale Integration). All use pipelining and/or multiple functional units to achieve vectorization/parallelization within each processor. Cray and IBM are the only supercomputer manufacturers to offer multiple-processors machines, although some additional vendors have announced multiprocessor machines. The form of synchronization on both the Cray and IBM machines is essentially event handling. Both Fujitsu and Hitachi systems are IBM System 370 compatible for scalar instructions.

2.2 Mini-Supercomputers

Below the supercomputer market, a new class of "near-supercomputers" or mini-supercomputers has emerged. These systems typically feature strong vector or advanced scalar capabilities and have been used for traditional high-performance technical computing applications. Priced well under supercomputers, $100,000 to generally no more than $1 million, mini-supercomputers are frequently sold when budgets are limited to this price range or when stand-alone (versus shared) capabilities are required. Early leaders in the field of mini-supercomputing were Alliant and Convex (see Table 2.3).

Table 2.3: **Mini-Supercomputers (performance in Mflops)**

Machine	Theoretical Peak Performance	First Shipment
Alliant FX/8	94	1985
Alliant FX/80	188	1987
Alliant FX/2800	1120	1990
Convex C-120	20	1984
Convex C-240	200	1987
DEC VAX 9000 440/VP	500	1990
FPS 500EA	668	1989
Cydrome †	25	1987
Gould CSD †	40	1987
Multiflow Trace 7/300 †	30	1988
SCS-40 †	44	1986

†Company no longer in business.

An alternative in the near-supercomputer category is the add-on array processor. Companies such as Floating Point Systems, Star Technology, and CSPI are actively marketing these add-on products in an effort to attract current supercomputer users.

2.3 Vector Mainframes

In a related vein, vector-processing enchancements are now being marketed for commercial mainframes. These vector enhancements allow machines produced for general-purpose applications to offer users increased numerical capability. In some cases, the ability to apply vectors is extended to more than one processor in multiprocessing mode. Companies currently offering such vector-processing capabilities include IBM, Control Data, Hitachi (NAS, COMPAREX), Unisys, and Honeywell. We summarize some of the machines in this category in Table 2.4.

Table 2.4: **Vector Mainframes**

Machine	Maximum Rate, in Mflops	Maximum Memory, in Mbytes	Maximum Number of Processors
CDC 180 995E	125	128	2
IBM 3090S/VF	800	256	1 - 6
NAS AS/EX VPF	484	64	1 - 4
Star ST-100	100	2048	4
Unisys 1190/ISP	266	128	1,2,4

2.4 Novel Parallel Processors

While most supercomputers and mini-supercomputers look to vector processing to provide performance, a number of companies are developing parallel-processing systems. Such systems range from smaller (8- to 32-processor) machines such as the Sequent or Encore to massively parallel (65,536-processor) systems such as the Thinking Machines CM-2.

Currently, two approaches to multiprocessing are being actively investigated: SIMD and MIMD. The three principal SIMD machines on the market are listed in Table 2.5.

Table 2.5: **SIMD Systems**

Company	System	Elements	Topology
Active Memory Tech.	DAP-610	1,024 - 4,096	nearest neighbor
MasPar	MP-1	1,024 - 16,384	nearest neighbor
Thinking Machines	CM-2	4,096 - 65,536	nearest neighbor/hypercube

MIMD systems fall into two classes: shared memory and distributed memory. Shared-memory architectures are composed of a varying number of processors and memory modules connected by means of a high-speed interconnect, such as a crossbar switch or efficient routing network. All processors share all memory modules and have the ability to execute different instructions on each of the processors by using different data streams. Examples of systems in this class of architecture are shown in Table 2.6.

Table 2.6: **MIMD Systems with Shared Memory**

Company	System	Elements	Connection
Alliant	FX/80	2 - 8	bus
Alliant	FX/2800	4 - 28	bus
BBN	TC/2000	2 - 256	butterfly network
Convex	C-200	2 - 4	bus
CRAY	2	1 - 4	crossbar
CRAY	Y-MP	1 - 8	crossbar
Elxsi †	6400	2 - 12	bus
Encore	Multimax	2 - 30	bus
Flex †	Flex/32	2 - 32	bus
Gould †	NP1	1 - 8	bus
IBM	3090J	1 - 6	crossbar
Myrias	(Canada)	2 - 512	network
NEC	SX-X	1 - 4	crossbar
Sequent	Symmetry	2 - 30	bus

†Company no longer in business

Distributed-memory architectures are composed of a number of processing nodes, each containing one or more processors, local memory, and communications interfaces to other nodes. Such architectures are scalable, have no shared memory among the processing nodes, exchange data through their network connections, and execute independent (multiple) instruction streams by using different data streams. The most popular architecture in this class is the hypercube. Examples of systems in this class are listed in Table 2.7.

Table 2.7: **MIMD Systems with Distributed Memory**

Company	System	Processors	Topology
FPS ‡	T-Series	2 - 16384	hypercube
Intel	iPSC/860	2 - 128	hypercube
Meiko	(England)	4 - 512	nearest neighbor
NCube	Model 2	64 - 2048	hypercube
Symult(Ametek) †	2010	2 - 256	nearest neighbor

†Company no longer in business
‡Company no longer manufacturing this product

In the discussion of parallel machines, the implementations of the two memory organizations are often misinterpreted. The term "shared memory" means that a single address space is accessible to every processor in the system. Hence any storage location may be read and written by any processor. Communication of concurrent processes is accomplished through *synchronized* access of shared variables in a shared memory system. The opposite of a shared address space should be multiple private address spaces, implying explicit communication. In this case communication between concurrent processes requires the explicit act of sending data from one processor to be received by another processor. These notions do not imply anything about the physical partitioning of the memory, nor anything about the proximity (in terms of access time) of a processor to a memory module. If physical memory is divided into modules with some placed near each processor (allowing faster access time to that memory), then physical memory is distributed. The real opposite of distributed memory is centralized memory, where access time to a physical memory location is the same for all processors. Hennessy and Patterson present a lively discussion of this topic [92].

From this line of reasoning, we can see that shared address vs. multiple addresses and distributed memory vs. centralized memory are separate, distinct issues. SIMD or MIMD architectures can have shared address and a distributed physical memory (for example, the BBN TC2000 and the Thinking Machines CM-2).

Figure 2.1 shows several machines according to this scheme.

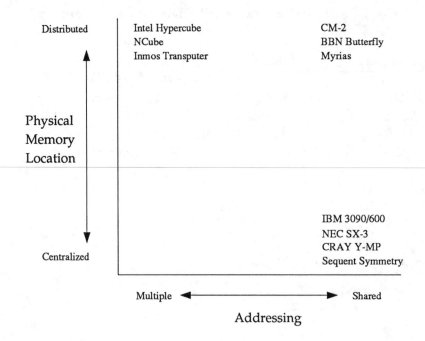

Figure 2.1: **Physical memory vs. address space**

In general, one can argue that it is easier to develop software for machines on the shared side of the addressing axis and that it is easier to build large-scale machines on the distributed end of the vertical axis.

It is our belief that parallel processing today is heading in the direction of a shared address space where the physical memory is distributed. In this model, the hardware deals with issues of data movement. This is not to say that data placement is not important. Indeed, from a standpoint of efficiency, data placement and minimization of data movement will be of paramount importance. No matter which communication scheme is used, the programmer and/or compiler must be conscious of the fact that the software generated will be for a large-scale parallel machine.

Since different memories have different access times, choices for placement and movement of data as the computation proceeds can have a significant effect on the overall performance of applications.

The model of programming from the user's point of view may look like that of Figure 2.2, processors in a *sea of memory*, where the processors have access to all memory through a shared address space, and the time to access the data depends on the distance the data is from the processor.

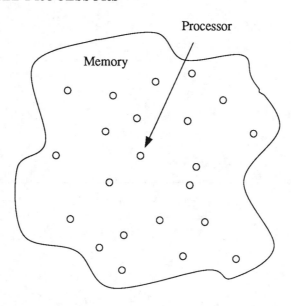

Figure 2.2: **Programming model from user's viewpoint**

Chapter 3

Implementation Details and Overhead

In this chapter, we are concerned with efficiency on vector and parallel processors. Specifically, we focus on implementation details that can decrease overhead and can improve performance. In a few instances we also discuss how an understanding of these implementation details can be exploited to achieve better efficiency.

3.1   Parallel Decomposition and Data Dependency Graphs

An illuminating way to describe a parallel computation is to use a control flow graph. Such a graph consists of nodes that represent processes and directed edges that represent execution dependencies. These graphs can represent both fine-grain and coarse-grain parallelism. However, as an aid to programming they are perhaps best restricted to the coarse-grain setting. In this case it is convenient to have the nodes represent serial subroutines or procedures that may execute in parallel. Leaves of such a graph represent processes that may execute immediately and in parallel. The graph asserts that there are no data dependencies between leaf processes. That is, there is no contention for write access to a common location between two leaf processes. The control flow representation of parallel computation differs from a data flow representation in the sense that the edges of the control flow graph do not represent data items. Instead, these edges represent assertions that data dependencies do not exist once there are no incoming edges. The programmer must take the responsibility for ensuring that the dependencies represented by the graph are valid.

The term *execution dependency* is self-explanatory. It means that one process is dependent

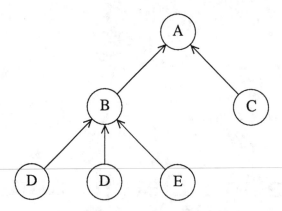

Figure 3.1: **Large-grain control flow graph**

upon the completion of, or output from, another process and is unable to begin execution until this condition is satisfied. Additional processes may in turn be dependent on the execution of this one. The parallel execution may be coordinated through the aggregation of this information into a control flow graph.

A special case of execution dependency is *deadlock*. This phenomenon, encountered only in MIMD machines, is caused by a circular dependency chain so that all activity ceases, with all processors either idle or in a wait state. Although it is sometimes described as the *bête noir* of parallel processing, it is by no means the most difficult bug to detect or fix. In particular, acyclic control flow graphs do not admit deadlock. Thus, describing a parallel decomposition in terms of such a graph avoids this problem from the outset.

We shall try to explain the concept of a control flow graph through a generic example (see Figure 3.1). In this example we show a parallel computation consisting of five processes. The nodes of the graph correspond to five subroutines (or procedures) A, B, C, D, and E (here with two "copies" of subroutine D operating on different data). We intend the execution to start simultaneously on subroutines C, D, D, and E since they appear as leaves in the dependency graph (D will be initiated twice with different data). Once D, D, and E have completed, B may execute. When B and C have completed execution, A may start. The entire computation is finished when A has completed.

This abstract representation of a parallel decomposition can readily be transformed into a parallel program. In fact, the transformation has great potential for automation. These ideas are developed more completely in [41, 42]. A nice feature of the control flow graph approach is that the level of detail required for the representation of a parallel program corresponds to the abstract level at which the programmer has partitioned his parallel algorithm. Let us illustrate this with a

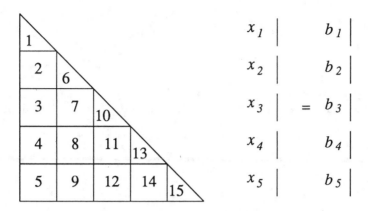

Figure 3.2: **Partitioning for the triangular solve**

simple example—the solution of a triangular linear system partitioned by blocks.

We can consider solving a triangular system of equations $Tx = b$ in parallel by partitioning the matrix T and vectors x and b as shown in Figure 3.2.

The first step is to solve the system $T_1 x_1 = b_1$. This will determine the solution for that part of the vector labeled x_1. After x_1 has been computed, it can be used to update the right-hand side with the computations

$$b_2 = b_2 - T_2 x_1$$

$$b_3 = b_3 - T_3 x_1$$

$$b_4 = b_4 - T_4 x_1$$

$$b_5 = b_5 - T_5 x_1.$$

Notice that these matrix-vector multiplications can occur in parallel, as there are no dependencies. However, there may be several processes attempting to update the value of a vector b_j (for example, 4, 8, and 11 will update b_4), and these will have to be synchronized through the use of locks or the use of temporary arrays for each process. As soon as b_2 has been updated, the computation of x_2 can proceed as $x_2 = T_6^{-1} b_2$. Notice that this computation is independent of the other matrix-vector operations involving b_3, b_4, and b_5. After x_2 has been computed, it can be used to update the right-hand side as follows:

$$b_3 = b_3 - T_7 x_2$$

$$b_4 = b_4 - T_8 x_2$$

$$b_5 = b_5 - T_9 x_2.$$

The process is continued until the full solution is determined. The control dependency graph for this is shown in Figure 3.3.

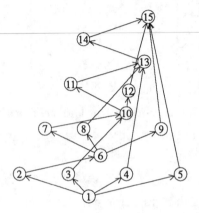

Figure 3.3: **Dependency graph for solution of triangular systems**

A complete understanding of the data dependencies among communicating parallel processes is, of course, essential to obtaining a correct parallel program. Once these dependencies have been understood, a control flow graph can be constructed that will assure that the data dependencies are satisfied. A control flow graph may go further by introducing execution dependencies to force a particular order of computation if that is deemed necessary for purposes of load balancing or performance.

3.2   Synchronization

Although the mechanisms for synchronization vary between machines, the necessity and reason for synchronization are the same in all cases. Synchronization is required when two or more concurrently executing processes must communicate, to signal their completion to each other, or to indicate that some data is now available for the use of the other, or simply to provide information to the other of its current progress. The two (or more) processes can then synchronize in the same way that people going on separate missions might synchronize their watches.

The crudest and most common form of synchronization is that of the *barrier*, often used in the join of a fork-and-join. Here a task records its completion and halts. Only when all (or a designated subset) of processes have reported to the barrier can the tasks halted at the barrier continue.

Most shared-memory parallel computers have at least one synchronization primitive. These are typically a test-and-set, or lock, function. A lock function tests the value of a lock variable. If the value is *on,* the process is blocked until the lock variable is *off.* When the lock variable is found to be *off,* it is read and set to *on* in an autonomous operation. Thus, exactly one process may acquire a lock variable and set it to *on.* A companion to a lock function is an unlock function. The purpose of an unlock function is to set the value of a lock variable to *off.*

These low-level synchronization primitives may be used to construct more elaborate primitives such as a barrier. However, explicit use of locks for synchronization should be avoided. Instead, one should try to use higher-level synchronization mechanisms available through the compiler or system, since the likelihood of introducing synchronization errors is high when low-level locks are used to construct intricate synchronization.

Locks are perhaps the easiest form of synchronization to use and understand. Normally, the programmer is allowed to designate locks by number. When a task wishes to have exclusive use of a section of code (and, presumably, exclusive use of the data accessed), it tries to set a lock through a subroutine call. In this situation a programmer has a section of program that is to be executed by at most one process at a time, called a *critical section.* If the lock has already been set by another task, it is unable to set the lock and continue into the critical section until the task that set that particular lock unsets or unlocks it. An example of the use of locks and critical sections follows:

```
CALL LOCK( <lock variable> )
     execute critical section
CALL UNLOCK( <lock variable> )
```

A more flexible and powerful form of synchronization can be achieved through the use of *events.* Events can be posted, awaited, checked for, and unposted. Any processor can unpost an event, and, after checking its existence, the programmer is free to continue if appropriate. Although this form of synchronization is not supported by all manufacturers, it is becoming increasingly common.

3.3 Load Balancing

Load balancing involves allocating tasks to processors so as to ensure the most efficient use of resources. With the advent of parallel processing, the ability to make iterations of loops run in parallel, referred to as microtasking, has become increasingly important. It can be very efficient to split off even small tasks: the determining factor is whether wall clock time has been reduced without having to pay significantly more for the increase in CPU time (resulting from parallel overhead).

When a job is being multiprocessed, a common measure of efficiency is the ratio of the elapsed time on one processor to that on p processors divided by p. This ratio reflects both time lost in communication or synchronization and idle time on any of the p processors. It is this idle time that load-balancing strategies seek to minimize.

To illustrate the effect of load balancing, let us assume that we have four processors and seven pieces of work (tasks) that can be executed simultaneously by any processor. Assume that six of the tasks each require 1 unit of work and the seventh requires 2 units. Clearly, a good load-balancing strategy would allocate the six jobs to three processors and let the longest task reside on the fourth. In that way the job can be completed in two units, the length of the longest task. If, however, two of the short tasks are executed on the same processor as the longest one, then the job will take 4 units and only half the processing power will be used.

The situation can become more complicated when the length of different tasks is not so simply related or if the length of a task is not known until it is executing. In a sense, we have a dynamic bin-filling problem to solve. We now—and not for the first time—encounter a tradeoff situation in parallel processing. Clearly, load balancing is easier if we have smaller tasks. However, smaller tasks have the adverse effect of increasing synchronization costs or startup overhead.

We acknowledge that the importance of load balancing *can* be overstated. For example, in a system that is multiprogrammed, the fact that one or more processors are idle should not be of great concern since the idle time (from one user's point of view) can be taken up by another job. This point is of particular importance when the parallelism, measured in number of simultaneously executable tasks, varies during the course of a job. Thus it might be efficient at one time to use all the processors of a system while at another time to use only one or two processors.

Finally, we observe that on a highly parallel machine, many tasks are required to saturate the machine, so that the problem size has to be large enough to generate enough parallelism. This phenomenon has been observed and documented by the group at Sandia (see, for example, [86, 87]) who have revised Amdahl's law on the basis of constant runtime rather than constant problem size (see Sections 4.1 and 4.3).

3.4 Recurrence

Recurrence refers to the use of a previously computed value of an expression used in a loop. One can have either a forward or a backward recurrence. Backward recurrences—where old values of earlier data are used—create no problems for vectorization or parallelism. Forward recurrences—the dependence of an assignment upon data that has just been reset—do inhibit vectorization. Consider the following simple loop:

```
      DO 10 I = 1,K
         A(I) = A(I+L)
10       CONTINUE
```

If the integer variable L is positive, we have a backward recurrence, and no problems with vectorization need arise. If L is negative, however, the new value for $A(I)$ is dependent on a value computed only L steps previously. This means that, without restructuring, the loop cannot be vectorized because the assignment to $A(I+L)$ will still be in the arithmetic pipes when the new value is needed for the assignment to $A(I)$. Finally, if the value of L is not known at compile time, then the compiler must assume that it is unsafe to vectorize and should not do so unless the programmer knows that its sign will always be positive and so informs the compiler by means of a directive.

Forward recurrences are often encountered in the solution of sparse linear systems, especially when iterative methods are used. Often, as in discretized two-dimensional and three-dimensional partial differential equations, these recurrences can be avoided by a suitable rearrangement of the computational scheme. For example, for certain classes of problem, a number of similar forward recurrences can be solved simultaneously, a procedure that then leads to possibilities for vectorization or parallelism. We shall see examples of this approach in Chapter 7.

In many situations, however, one really needs to solve only one (large) bi- or tridiagonal system before other computations can proceed. Three basic techniques exist for vectorizing or parallelizing the algorithm for these linear systems: recursive doubling [46, 154], cyclic reduction [91, 113], and divide and conquer [21, 170]. Let us briefly describe these three techniques with respect to their application for bidiagonal systems (the application to tridiagonal systems is similar and is described in the references). For simplicity we assume that the bidiagonal system has a unit diagonal (this can often be arranged in iterative solution methods, and it saves costly division operations; see Chapter 7). Then the solution of

$$\begin{pmatrix} 1 & & & & & \\ -a_2 & 1 & & & & \\ & -a_3 & 1 & & & \\ & & \cdot & \cdot & & \\ & & & \cdot & \cdot & \\ & & & & \cdot & \\ & & & -a_n & 1 \end{pmatrix} \begin{pmatrix} x_1 \\ x_2 \\ x_3 \\ \cdot \\ \cdot \\ \cdot \\ x_n \end{pmatrix} = \begin{pmatrix} b_1 \\ b_2 \\ b_3 \\ \cdot \\ \cdot \\ \cdot \\ b_n \end{pmatrix}$$

is obtained by the forward recurrence

$$x_1 = b_1$$

$$x_i = b_i + a_i * x_{i-1}, i = 2, 3, ..., n.$$

The basic trick of *recursive doubling* is to insert the recursion for x_{i-1} in the right-hand side:

$$\begin{aligned} x_i &= b_i + a_i * (b_{i-1} + a_{i-1} * x_{i-2}) \\ &= \tilde{b}_i + \tilde{a}_i * x_{i-2}. \end{aligned}$$

Hence we obtain two recurrence relations: one for the even indexed unknowns and one for the odd indexed unknowns. These relations have been obtained at the cost of the computation of the new coefficients $\tilde{a}_i$ and $\tilde{b}_i$, which can be done in vector mode or in parallel. The trick can be repeated and leads each time to twice as many recurrence relations of half the length of the previous step. For some parallel architectures this may be an advantage, but for vector machines it does not really relieve the problem.

The situation is slightly better with *cyclic reduction*, where the repetition of the recurrence relation is carried out only for the even indexed unknowns. This leaves a recurrence of half the original length for the even unknowns. After that has been solved, the odd indexed unknowns can be solved by one vector operation. This trick can also be repeated, and many optimized codes are based on a few steps of cyclic reduction. On vector machines it is often advantageous to combine two successive steps of cyclic reduction.

The third technique can be regarded as a *divide and conquer* approach. The approach is to view the given bidiagonal system as a block bidiagonal system. In the first step the off-diagonal elements in all the diagonal blocks are eliminated, which can be done in parallel or in vector mode.

This step leads to fill-in in the off-diagonal blocks. The fill-in is then eliminated in the second step, which can also be done either in parallel or in vector mode. The divide and conquer technique was first proposed in [21], but it is more widely known as Wang's method after Wang [170] who described the method for the parallel solution of tridiagonal systems. In [166] this approach is taken for the vectorization of the algorithm for bidiagonal systems, and its performance is analyzed for some different vector architectures. The divide and conquer technique has been generalized by Meier [123] and Dongarra and Johnsson [39] for more general banded linear systems.

All three techniques lead to a substantial increase in the number of floating-point operations (for bidiagonal systems, a factor of about 2.5), which effectively reduces the gains obtained by vectorization or parallelism. This makes these techniques unattractive in a classical serial environment. Another potential disadvantage is that these techniques require the systems to be of huge order (several thousands of unknowns) in order to outperform the standard forward recurrence with respect to CPU time. Often, scalar optimization by loop unrolling leads to surprisingly effective code. See [141] for an extreme example.

3.5 Indirect Addressing

Simple loops involving operations on full vectors can be implemented efficiently on both vector and parallel machines, particularly if access is to components stored contiguously in memory. For many machines, constant stride vectors can be handled with almost the same efficiency so long as bank conflict or cache problems are avoided. For many calculations, however—for example, computations on sparse matrices—access is not at all regular and may be directed by a separate index vector.

A typical example is a sparse SAXPY operation of the form

```
     DO 10 I = 1,K
        W(I) = W(I) + ALPHA * A(ICN(I))
  10 CONTINUE
```

where the entries of A are chosen by using the index vector ICN. One problem is that the compiler typically will have no knowledge of the values in ICN and hence may not be able to reorganize the loop for good vectorization or parallelism. The original vector supercomputers, notably the CRAY-1, could do nothing but execute scalar code with such loops, but now most machines offer hardware indirect addressing which allows vectorization.

The implementation varies from machine to machine. For a vector supercomputer, the entries of ICN are usually loaded into a vector register and then used in a hardware instruction to access the relevant entries of A. If the machine has several pipes to memory, a high asymptotic rate can still be reached, although the startup time is usually substantially greater than for similar operations with direct addressing because of the extra loads for ICN. We see that typically between a quarter and a half of the asymptotic rate for directly addressed loops can be obtained with a startup time of about twice the directly addressed counterpart.

The moral is that today we need not shy totally away from indirect addressing, although we should be aware of the possibilities of bank conflict and the higher startup time. The last factor can be especially important when dealing with sparse matrices because typical loop lengths are related to the number of entries in a row or column rather than the problem dimension. This situation has led, for example, to novel schemes for holding sparse matrices prior to matrix-vector multiplication (see, for example, Saad [138] or Radicati and Robert [134]) in order to increase the length of the indirectly addressed loop.

A final word of advice: if one is assigning to an indirectly addressed vector (that is, a reference of the form A(ICN(I)) appears on the left of the equation), an automatic parallelizer cannot know whether repeated reference is being made to the same location, and so automatic parallelizing will not take place. In such instances, the programmer may choose to use compiler directives to ensure parallelization of the loop.

Chapter 4

Performance: Analysis, Modeling, and Measurements

The computational speed of modern architectures is usually expressed in terms of Mflops. For the most powerful of today's supercomputers, even Gigaflops are used to describe their high-performance (1 Gflop = 1000 Mflops). From the definition of Mflops it follows that when N floating-point operations (flops) are carried out in t microseconds (10^{-6} seconds), the speed of computation is given by

$$r = \frac{N}{t} \text{ Mflops},\qquad\qquad (4.1)$$

and, vice versa, when N flops are executed with an average computing speed of r Mflops, then the CPU time is

$$t = \frac{N}{r} \text{ microsec.}\qquad\qquad (4.2)$$

Different operations may be executed at quite different speeds, depending, for example, on the advantage taken of pipelining possibilities, references to memory, etc. This fact implies that we cannot simply analyze the performance of an algorithm by counting the number of flops to be executed (as used to be a good indication for classical computers). Instead, we must take into account the fact that specific parts of an algorithm may exploit the opportunities offered by a given architecture. In this chapter, we describe the principles for analyzing and modeling the performance of computational work to help predict, understand, and guide the software/program development.

4.1   Amdahl's Law

In most realistic situations not all operations in a complex algorithm are executable with high computational speeds. When two parts of a job are executed each at a different speed, the total CPU time can be expressed as a function of these speeds. It turns out that the lower of these speeds may have a dramatic influence on the overall performance. This effect was first pointed out by Amdahl [4], and therefore the relation between performance (computational speed) and CPU time is often called *Amdahl's law*.

4.1.1   Simple Case of Amdahl's Law

We first discuss a simple case of Amdahl's law. Let us assume that the execution of a given algorithm involves N flops and that on a certain computer a fraction f of these flops is carried out with a speed of V Mflops, while the rest is executed at a rate of S Mflops. Let us further assume that this fraction f is well suited to the given architecture, while the other part does not take much advantage of speedup possibilities ($V \gg S$). The total CPU time t, according to (4.2), is expressed by

$$t = \frac{fN}{V} + \frac{(1-f)N}{S} = N(\frac{f}{V} + \frac{1-f}{S}) \text{ microsec,} \tag{4.3}$$

and for the overall performance r we have from (4.1) that

$$r = \frac{N}{t} = \frac{1}{f/V + (1-f)/S} \text{ Mflops (Amdahl's law).} \tag{4.4}$$

From (4.3) we see that

$$t > \frac{(1-f)N}{S} \text{ microsec.} \tag{4.5}$$

If the algorithm had executed completely with the lower speed S, the CPU time would have been $t = N/S$ microseconds. This implies that the relative gain in CPU time obtained from executing the portion fN at the speed V instead of the much lower speed S is bounded by $\frac{1}{1-f}$. Thus, f must be rather close to 1 in order to benefit significantly from high computational speeds.

This effect is nicely illustrated by Figure 4.1, where we display the overall performance r as a function of f for the situation in which $V = 130$ Mflops and $S = 4$ Mflops. These numbers represent a realistic situation for the CRAY-1 (on more recent computers the ratio $\frac{V}{S}$ can be much larger, depending on the type of operation).

We see that f has to be quite large—say, larger than 0.8 (or 80%)—in order to obtain significant improvements in the overall speed. Since we usually encounter a whole range of speeds for the

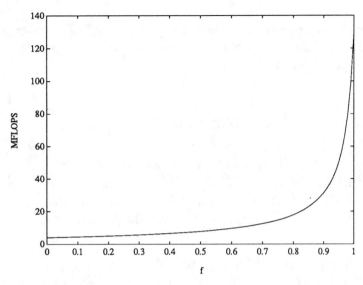

Figure 4.1: **Amdahl's law for the performance** r

different parts of an algorithm, we also give a more general form of Amdahl's law in order to facilitate the analysis of realistic situations.

4.1.2 General Form of Amdahl's Law

Assume that the execution of an algorithm consists of executing the consecutive parts $A_1, A_2, ..., A_n$ such that the N_j flops of part A_j are executed at a speed of r_j Mflops. Then the overall performance r for the $N = N_1 + N_2 + ... + N_n$ flops is given by

$$r = \frac{N}{(N_1/r_1 + N_2/r_2 + ... + N_n/r_n)} \text{ Mflops.} \tag{4.6}$$

This follows directly from the observation that the CPU time t_j for the execution of part A_j is given by $t_j = \frac{N_j}{r_j}$ microsec. Hence the total time for the algorithm is the sum of these times, and (4.1) gives the desired result.

With Amdahl's law we can study the effect of different speeds for separate parts of a computational task. However, this is not the end of the story. Previously we have seen that improvements in the execution of computations usually are obtained only after some startup time. This fact implies that the computational speed depends strongly on the amount (and type) of work. In Section 4.2 we shall discuss this effect for pipelined operations.

4.2   Vector Speed and Vector Length

We have seen in Section 1.2.2 that operations that can be pipelined—the so-called vector operations–require a certain startup time before the first results are delivered. Once the pipeline produces results, the subsequent results are delivered at much shorter intervals. Hence the average computing speed for long vectors may be rather large, but for very short vectors the effect of pipelining may be barely noticeable.

The performance r_N for a given loop of length N is usually expressed, after Hockney and Jesshope [94], in terms of parameters r_∞ and $n_{1/2}$. The values of these parameters are characteristic for a specific loop: they represent the performance in Mflops for very long loop length (r_∞), and the loop length for which a performance of about $1/2r_\infty$ Mflops is achieved ($n_{1/2}$).
For many situations we have in fairly good approximation that

$$ r_N = \frac{r_\infty}{n_{1/2}/N + 1} \text{ Mflops.} \tag{4.7} $$

The actually observed Mflops rates may differ slightly from the model, since (4.7) does not account for stripmining effects (see Section 1.2.10) or memory bank conflicts (see Section 1.3.1).

It is straightforward from (4.7) that for loop length $N = qn_{1/2}$ a performance of $r_\infty q/(1 + q)$ Mflops will be obtained. Consequently, if in a given application the loop length is N, then it is highly desirable that the $n_{1/2}$-values concerned be small relative to N.

As a rule of thumb we have that $n_{1/2}$ and r_∞ are proportional to the number of pipelines of a given computer. If a model comes with different numbers of pipelines, then of course the r_∞-values for the largest model are highest, but it will also require larger vector lengths to obtain reasonable performances.

The same holds on a p-processor parallel computer when parallelism in a code is realized through splitting the loops over the processors. For many loops this approach will lead to an increase of r_∞ and $n_{1/2}$ by a factor of p relative to single-processor execution. In general, however, on a p-processor system r_∞ grows less than with p, and $n_{1/2}$ grows faster, because of synchronization effects. Especially in sparse matrix codes, parallelism is often obtained through loop splitting; and hence we must have fairly large vector lengths for high-performance execution on a parallel multiple-pipeline processor system (as, e.g., the NEC SX-3).

Our modeling approach for the performance of parallel vector systems has been simplified in that we have not taken into account the effects of memory hierarchy (e.g., cache memory). For computational models that include these effects see, e.g., [74, 95].

4.3 Amdahl's Law—Parallel Processing

Pipelining is only one way to help improve performance. Another way is to execute parts of a job in parallel on different processors. Of course, these processors themselves may be vector processors, a situation that further complicates the analysis.

4.3.1 A Simple Model

In parallel processing the total CPU time is usually larger than the CPU time involved if the computation had taken place on a single processor, and it is much larger than the CPU time spent per processor. Since the goal of parallel processing is to reduce the wall clock time, we shall compare the wall clock times for different numbers of processors in order to study the accelerating effect of parallel processing.

It may be expected that if the computations can be carried out completely in p equal parallel parts, the wall clock time will be nearly $1/p$ of that for execution on only one processor. As is already clear from Amdahl's law, the non-parallel (or serial) parts may have a negative influence on this reduction. We shall make this more explicit.

Let t_j denote the wall clock time to execute a given job on j parallel processors. The speedup, S_p, for a system of p processors is then by definition

$$S_p = \frac{t_1}{t_p}. \tag{4.8}$$

The efficiency, E_p, of the p-processor system is defined by

$$E_p = \frac{S_p}{p}. \tag{4.9}$$

The value of E_p expresses, in a relative sense, how busy the processors are kept. Evidently $0 \leq E_p \leq 1$.

Let us assume for simplicity that a job consists of basic operations all carried out with the same computational speed and that a fraction f of these operations can be carried out in parallel on p processors, while the rest of the work can keep only 1 processor busy. The wall clock time for the parallel part is then given by $\frac{ft_1}{p}$; and, if we ignore all synchronization overhead, the time for the serial part is given by $(1 - f)t_1$. Consequently, the total time for the p-processor system is

$$t_p = \frac{ft_1}{p} + (1 - f)t_1 = \frac{t_1(f + (1 - f)p)}{p} \geq (1 - f)t_1. \tag{4.10}$$

From this the speedup S_p directly follows:

$$S_p = \frac{p}{(f + (1 - f)p)}. \tag{4.11}$$

Note that $f < 1$ and the speedup S_p is bounded by $S_p \leq \frac{1}{1-f}$. Relation (4.11) is often referred to as *Ware's law* [171].

These relations show that the speedup S_p is reduced by a factor $f + (1 - f)p = O(p)$, which leads to considerable reductions for increasing values of p, even for f very close to 1. For example, when 90% of the work can be done entirely in parallel on 10 processors, then the speedup is only about 5 (instead of the desired value 10), and the speedup for very large p is limited by 10.

The model (4.11) represents a highly idealized and simplified situation. In practice, matters are usually complicated by factors such as the following:

- The number of parallel portions of work has to be large enough to match a given number of processors.

- The (negative) influence of synchronization overhead for the parallel execution has been ignored.

- Often one has to modify part of the underlying algorithm in order to obtain a sufficient degree of parallelism, a procedure that in many situations leads to increasing computational complexity.

Nevertheless, the simple model is useful since it helps us to predict upper bounds for the speedup. In Table 4.1 we list the speedup for several values of the parallelizable fraction f and the number of processors p, under the assumption that the parallel fraction can be executed entirely in parallel, without any overhead, on p processors.

Table 4.1: **Speedups for Values of f and p**

f	$p=1$	$p=2$	$p=3$	$p=4$	$p=8$	$p=16$	$p=32$	$p=64$	$p=\infty$
1.00	1.00	2.00	3.00	4.00	8.00	16.00	32.00	64.00	∞
.99	1.00	1.98	2.94	3.88	7.48	13.91	24.43	39.26	100.00
.98	1.00	1.96	2.88	3.77	7.02	12.31	19.75	28.32	50.00
.96	1.00	1.92	2.78	3.57	6.25	10.00	14.29	18.18	25.00
.94	1.00	1.89	2.68	3.39	5.63	8.42	11.19	13.39	16.67
.92	1.00	1.85	2.59	3.23	5.13	7.27	9.19	10.60	12.50
.90	1.00	1.82	2.50	3.08	4.71	6.40	7.80	8.77	10.00
.75	1.00	1.60	2.00	2.28	2.91	3.37	3.66	3.82	4.00
.50	1.00	1.33	1.50	1.60	1.78	1.88	1.94	1.97	2.00
.25	1.00	1.14	1.20	1.23	1.28	1.31	1.32	1.33	1.33
.10	1.00	1.05	1.07	1.08	1.09	1.10	1.11	1.11	1.11
.00	1.00	1.00	1.00	1.00	1.00	1.00	1.00	1.00	1.00

For example, suppose that we have observed a speedup $S_4 = 3.77$ on a four-processor system for a certain job. Assume that this job can be seen as being composed of a serial part and a parallel part, suitable for nicely distributed parallel processing on at least four processors. It follows from Table 4.1 (and the equation (4.11)) that the fraction of parallel work is about 98%. This does not look too bad, but if this job could also be executed (without any additional overhead or complications) on 16 processors, then the speedup would have been 12.31 at most.

In fact, we conclude from Table 4.1 that even for this apparently highly parallel job, the speedup soon falls well behind the number of processors. In most practical situations it is quite unrealistic to assume that a given job consists of only two parts, a strictly serial part and one that can execute in parallel on p processors. Often, some parts can be executed in parallel on p processors while other parts can still be executed in parallel but on fewer than p processors.

Moler [131] properly observes that an increase in the number of processors usually goes along with an increase in the problem size, and he introduces the notion of being *"effectively parallel"* to go along with this. (See also Gustafson's model in Section 4.3.1.) Let the non-parallel part of the computations be denoted by $\alpha(M) = 1 - f$, where M is a measure of the total number of operations. Moler reports that for many algorithms, $\alpha(M)$ decreases when M is increased. In line with this, an "effectively parallel algorithm" is defined as an algorithm for which $\alpha(M)$ goes to zero when M grows.

If the parallel part can be executed in parallel on q processors, then it follows for such an algorithm that

$$S_q = \frac{q}{1 + (q-1)a(M)} \quad \text{and} \quad E_q = \frac{1}{1 + (q-1)a(M)} \tag{4.12}$$

and hence S_q grows to q and E_q grows to 1 for growing M.

Now the question arises whether (local) memory is large enough to handle a problem size for which S_q is sufficiently close to q. Moler notes that in many situations memory size is a bottleneck in realizing large speedups, simply because it often limits the problem size.

4.3.2   Gustafson's Model

The basic premise of Amdahl's law is that the proportion of code that can be vectorized or parallelized remains constant as vector length (that is, problem size) increases. As Gustafson [86] pointed out, this is often not appropriate. An alternative model can be based on fixing the computation time rather than the problem size, as f varies. Thus, in Gustafson's model, if we assume that the problem can be solved in one unit of time on our parallel machine and that the time in serial computation is $1 - f$ and in parallel computation is f, then if there are p processors in our parallel system, the time for a uniprocessor to perform the same job would be $1 - f + fp$. Then the speedup, $S_{p,f}$, is given by

$$S_{p,f} = p + (1 - p)(1 - f).$$

We see that, as the number of processors increases, so does the proportion of work in parallel mode, and speedup is not limited in the same way as in Amdahl's law. By using this measure, the Sandia team [87] obtained near-optimal speedups on an NCube machine to win the first Bell Prize in 1988. (Gordon Bell established an award in 1987 to demonstrate his personal commitment to parallel processing. The prize consists of two \$1,000 awards to be given annually for ten years for the best scientific or engineering program with the greatest speedup [40].)

4.4   Examples of $(r_\infty, n_{1/2})$-values for Various Computers

As we have seen in the preceding sections, the performance for a given (vector) operation depends on the architecture of the computer and on the ability of its Fortran compiler to generate efficient code. From a modeling point of view the length of the clock cycle is of no importance, since it is only a scaling factor for r_∞. The factors that really make computers behave differently include features such as direct memory access, numbers of paths to memory, the bandwidth, memory hierarchy, and chaining.

To give some idea of the impact of these architectural aspects on the performance, we report on our own measurements of the performance parameters for widely different architectures. In this section, we consider rather simple Fortran DO-loops, for which we measured the CPU times for different vector lengths (see Table 4.2). By a simple linear least-squares approximation, we determine the values of r_∞ and $n_{1/2}$ for a number of supercomputers and mini-supercomputers. In Section 4.5, we shall present performance measurements for a more complicated algorithm (taken from LINPACK).

Table 4.2: **Various Simple Fortran DO-Loops**

Number of the Loop	Number of Flops per Pass (n)	Operation per Pass of the Loop	Operation
1	2	v1(i)=v1(i)+a*v2(i)	update
2	8	v1(i)=v1(i)+s1*v2(i)+s2*v3(i)+ s3*v4(i)+ s4*v5(i)	4-fold vector update
3	1	v1(i)=v2(i)/v3(i)	divide
4	2	v1(i)=v1(i)+s*v2(ind(i))	update+gather
5	2	v1(i)=v2(i)-v3(i)*v1(i-1)	bidiagonal system
6	2	s=s+v1(i)*v2(i)	innerproduct
7	9	v1(i)=v2(i)*v0(i)+v3(i)*v0(i-1)+ v4(i)*v0(i+1)+v5(i)*v0(i-m)+ v6(i)*v0(i+m)	sparse matrix vector

We have deliberately restricted ourselves to this special selection because they play a role in our algorithms:

- Loop 1 is the well-known vector update, which is central in standard Gaussian elimination and in updating of current iteration vectors in iterative processes. Therefore we will see this loop again in Chapters 5 and 7.

- Loop 2 represents the combined effect of 4 vector updates, often referred to as a GAXPY. The performances for this loop, compared to those for Loop 1, show how an architecture (or compiler) can improve performance by, e.g., keeping $v1$ in the registers. This capability is fully exploited in the so-called Level 2 BLAS approach (Chapter 5). Hence, if the performances for Loop 2 are markedly better than those for Loop 1, we may expect big improvements by using Level 2 BLAS kernels instead of Level 1 BLAS kernels (this point will be further explained and highlighted in Chapter 5).

- Loop 3 has been included to demonstrate how slow a full-precision divide operation can be. Divide operations, which occur quite naturally in matrix computations (e.g., scaling), often can be largely replaced by multiply operations with the only once-computed inverse elements.

- Loop 4 is quite representative for direct solvers for sparse matrix systems. It also indicates the speed-reducing effects of indirect addressing, which is also typical of sparse matrix computations. Loops of this type will be encountered in Chapter 6.

- Loop 5 is often seen in the solution of (block) tridiagonal systems, as well as in some preconditioning operations (this will be explained in Chapter 7). It also shows to some extent how slow scalar operations on a vector computer can be.

- Loop 6 is frequently used in iterative processes for the computation of iteration parameters and for stopping criteria (Chapter 7).

- Loop 7 models the matrix vector product for the two-dimensional discretized Laplacian and may very well occur as a kernel in iterative solvers for such systems. We shall see such operations more than once in Chapter 7.

The reader who wishes to determine the $(r_\infty, n_{1/2})$-values for other loops, or in other circumstances, may do so by using a code that we have made available in *netlib* (see Appendix A).

4.4.1   CRAY-1 and CRAY-2 (one processor)

Table 4.3: **Performance of CRAY-1 and CRAY-2**

Loop Number	CRAY-1 CFT77 V1.3 (Feb. 1988)		CRAY-2 (1 proc.) CFT77 V2.0 (Apr. 1988)	
	r_∞	$n_{1/2}$	r_∞	$n_{1/2}$
1	46	10	71	41
2	77	8	110	16
3	13.6	9	24	33
4	6.3	12	38	27
5	10	2	8.5	10
6	75	179	88	184
7	54	7	77	23

Knowing only the clock cycle times, one would expect the CRAY-2 to be about three times faster than the CRAY-1. We see in Table 4.3 that this is not the case, except for indirect addressing constructions (these are slow on the CRAY-1 since this machine does not have hardware gather/scatter instructions). The main reason that the CRAY-2 is not that fast (in Fortran!) is that the hardware does not allow chaining of operations. It is, of course, possible to write more efficient code in Cray assembly language (CAL) for many of the above operations. For example, for the innerproduct operation (Loop 6), code has been written that attains an r_∞ greater than 200 Mflops for the CRAY-2 [43].

4.4.2 CRAY X-MP (one processor; clock cycle time 8.5 nsec)

To demonstrate the progress in compilers and the dependence of performance measurements on the quality of the Fortran compilers, we give performance parameters for two different compilers on the CRAY X-MP.

Table 4.4: **Performance of CRAY X-MP (one processor)**

Loop Number	CFT 1.15 BF2		CFT77 V2.0*207	
	r_∞	$n_{1/2}$	r_∞	$n_{1/2}$
1	166	84	166	29
2	202	30	207	13
3	26	24	31	16
4	82	44	94	19
5	5.7	1	15.3	13
6	166	292	206	286
7	178	27	174	16

We see from Table 4.4 that the r_∞ value has improved for several loops, but most impressive is the reduction of the $n_{1/2}$ value for most loops. This means that real supercomputing speeds are often obtained for very modest vector lengths N. Other remarkable points are that the CRAY X-MP architecture leads to a performance close to the peak performance (235 Mflops) in Fortran for some important loops, in contrast with the CRAY-1 or CRAY-2 for which it is not so easy to approach peak performance. Note also that indirect addressing leads to a reduction in speed only by a factor of roughly two (this also in contrast with the CRAY-1 and many other supercomputers).

4.4.3 CYBER 205 (2-pipe) and ETA-10P (single processor)

The CYBER 205 computers were superseded by the ETA-10 computers. It is therefore of interest to see how the designers of the architecture attempted to improve on the older 205 models. In comparing the data (see Table 4.5), one must realize that clock cycles are different for the tested machines (CYBER 205: 20 nsec, ETA-10P: 24 nsec), so their r_∞ values must be compared in a relative sense. Both machines have been tested with the same Fortran compiler: FORTRAN200 - level 678.

Table 4.5: **Performance of CYBER 205 and ETA-10P**

Loop Number	CYBER 205 r_∞	$n_{1/2}$	ETA-10P r_∞	$n_{1/2}$
1	200	170	167	140
2	200	160	167	55
3	16	20	13.6	21
4	30	46	44	86
5	2.9	1	1.6	1
6	100	162	83.3	243
7	100	96	83.3	70

For both machines the peak performance is actually achieved for some loops, because of lack of stripmining effects. Note the more favorable performance of the ETA-10P with respect to indirect addressing.

For very simple loops, such as Loop 6, we see that the scalar loop overhead leads to a rather large $n_{1/2}$. Actually, performance results as represented above might inspire us to rather misleading conclusions, since for more complicated problems, such as Loop 2, we see that the ETA-10P outperforms the CYBER 205 even for smaller vector lengths. The explanation is that for more complicated loops (which comprise more vector operations), the scalar overhead is much less than the sum of the overheads for each of the individual vector operations. Furthermore, there seems to be more opportunity to overlap this scalar overhead with the vector operations.

4.4.4 IBM 3090/VF (1 processor; clock cycle time 18.5 nsec)

Each processor of the IBM 3090 has a relatively small cache memory, 64 kilobytes. The performance parameters were determined for vector operands that were not available in cache at the moment of starting the vector operations. The timing experiments reported in Table 4.6 were carried out with the Fortran compiler FORTVS2 (as of December 10, 1988).

Table 4.6: **Performance of IBM 3090/VF (1 processor)**

Loop Number	r_∞	$n_{1/2}$	$r_{\max}(cache)$
1	21	46	30
2	41	42	61
3	2.7	16	2.9
4	11	34	16
5	6.5	8	8.8
6	29	74	45
7	27	37	38

In many dense matrix algorithms, it is possible to reuse intermediate data stored in cache memory. In that case, the performance may be much better, as is shown in the right half of Table 4.6 in the column labeled $r_{\max}(cache)$. This column gives the maximum performances for operations for which all the operands happen to be in cache. The size of the cache and the type of operation determine the maximum vector length of the operations; for example, for the vector update, the maximum vector length is about 1500, while for the sparse matrix-vector product, the maximum vector length is about 500 (with the size of cache on this machine). For sparse matrix computations it will, in general, be difficult to use the cache efficiently, especially when indirect addressing is involved.

4.4.5   NEC SX/2

The NEC SX/2 has a cache memory for scalar operands. This hardly influences the performance for most vector operations. Also in this case we have assumed that, at the start of the vector loop, scalar data such as addressing information was not available in cache. The Fortran compiler FORTRAN 77/SX, Rev. 036, has been used for the timing experiments.

Table 4.7: **Performance on the NEC SX/2**

Loop Number	r_∞	$n_{1/2}$
1	548	148
2	900	108
3	61	50
4	35	30
5	21	8
6	905	607
7	843	106

As Table 4.7 shows, the performance is relatively poor when indirect addressing is involved; the performance for the update is reduced by a factor of about 16.

Note also the rather modest performance for the vector divide. For the vector-vector multiply the values are $r_\infty = 270$ Mflops and $n_{1/2} = 151$. Clearly, it pays to invert the vectors if they are required more than once in vector-divide operations.

4.4.6   Convex C-1 and Convex C-210

The Convex C-1 and C-210 computers represent two generations of mini-supercomputers and therefore can be used to show the progress that has been made. Both machines were tested under the same Fortran 77 compiler (as of December 9, 1988).

Table 4.8: **Performance of the Convex C-1 and C-210**

Loop Number	Convex C-1		Convex C-210	
	r_∞	$n_{1/2}$	r_∞	$n_{1/2}$
1	6.0	14	16	26
2	10.7	22	23	15
3	1.8	13	4.0	1
4	4.0	12	9.0	20
5	.87	1	1.5	1
6	5.9	28	18	36
7	7.0	17	13	22

The r_∞ values for the C-210 (shown in Table 4.8) nicely reflect the reduced cycle time c with respect to the C-1 ($c = 100$ nanoseconds for the C-1 and $c = 40$ nanoseconds for the C-210).

4.4.7   Alliant FX/80

An Alliant FX/80 may be equipped with up to 8 processors (ACEs). Table 4.9 gives timing results collected for different numbers of ACEs. The compiler FORTRAN V3.1.33 (D.20D52) was used for all the timings.

Table 4.9: **Performance of Alliant FX/80**

Loop Number	1 ACE		2 ACEs		4 ACEs		6 ACEs	
	r_∞	$n_{1/2}$	r_∞	$n_{1/2}$	r_∞	$n_{1/2}$	r_∞	$n_{1/2}$
1	3.45	18	6.9	55	13.8	96	17.1	117
2	5.6	6	12.2	25	23.0	47	32.6	74
3	1.1	11	2.2	29	4.3	59	6.4	94
4	2.8	15	4.5	28	9.75	71	10.85	77
5	1.6	22	1.6	22	2.14	43	3.0	87
6	5.1	31	10.2	78	20.4	160	30.5	254
7	3.6	5	6.2	8	13.3	26	16.8	31

Before analyzing the results, we note the following points:

- The timings for 1 ACE have been obtained with other compiler options, including switching parallelism off to make the code for 1 ACE a little more efficient. The timings for multiple ACEs have been obtained by explicitly parallel code. Hence, it becomes difficult to compare the results for 1 ACE with those for multiple ACEs.

- No special care has been taken to let the loop lengths be exact multiples of the numbers of ACEs. Occasionally, therefore, the performance is slightly degraded; and hence the least-squares fit for r_∞ leads to slightly less than optimal values.

- Because of the cache, we sometimes experience degradation in performance, especially for relatively simple loops (see the performance for the LINPACK code in Section 4.5.3). For more complicated DO-loops (e.g., Loop 2), the cache can be better exploited. This may partially explain irregularities in the performance parameters for multiple ACEs.

From the results in Table 4.9 we conclude that, indeed, in many cases r_∞ is proportional to the number of ACEs. We also see that, at least for some loops, the $n_{1/2}$-values for multiple ACEs grow roughly linearly with the number of ACEs (e.g., Loops 2, 3, and 6). In those cases the synchronization overhead apparently is small.

Finally, we note that the performance for combined updates, as in Loop 2, is much better than for a single update (Loop 1); hence the Level 2 BLAS approach (Section 5.1.2) is an improvement. A closer look at our experiments reveals that the performance for such combined loops is even better for loop lengths that fit the cache size; this can be realized through the use of the Level 3 BLAS (Section 5.1.3).

4.4.8   General Observations

From the preceding tables we conclude that, for most architectures, the simple DO-loops do not lead to anything close to peak performance. Only those machines capable of moving at least three vectors concurrently to and from main memory achieve peak performance for some of the loops (e.g., CRAY X-MP, CRAY Y-MP, ETA, and CYBER).

An additional bottleneck is caused by memory hierarchy that involves a cache (e.g., on the IBM and Alliant systems). This leads to a further degradation in the actual performance in many cases.

One should not conclude from our findings that it is impossible to come close to peak performance for important algorithms on all those machines. In fact, our results indicate that we have to look for possibilities of combining loops, in order to save on memory transfer. We discuss loop combining further in Chapter 5, where we show what can be done by using a set of operations over two and three loops.

We note, however, that in sparse matrix applications we are often restricted in the possibilities for loop combining. Then the $(r_\infty, n_{1/2})$-parameters may help to understand or model the performance of more complex algorithms.

4.5   LINPACK Benchmark

Simple DO-loops, like those in Section 4.4, form the backbone for the LINPACK subroutines. We therefore include some discussion on how such more complicated subroutines behave on vector and parallel architectures.

As we shall see, subroutines that merely use vector operations provide hardly any improvement over a simple DO-loop when executing on a machine with a cache memory. To avoid the bottlenecks inherent in such a memory hierarchy, we need the Level 2 or Level 3 BLAS approach, which will be treated in detail in Chapter 5.

4.5.1   Description of the Benchmark

Before continuing, we note here the distinction between LINPACK [30] and the so-called LINPACK benchmark. LINPACK is a collection of subroutines to solve various systems of simultaneous linear algebraic equations. The package was designed to be machine independent and fully portable and to run efficiently in many operating environments. The LINPACK benchmark [29] measures the performance of two routines from the LINPACK collection of software. These routines are

DGEFA and DGESL (these are double-precision versions; SGEFA and SGESL are their single-precision counterparts).

DGEFA performs the LU decomposition with partial pivoting, and DGESL uses that decomposition to solve the given system of linear equations. Most of the time is spent in DGEFA. Once the matrix has been decomposed, DGESL is used to find the solution; this process requires $O(n^2)$ floating-point operations, as opposed to the $O(n^3)$ floating-point operations of DGEFA.

4.5.2   Calls to the BLAS

DGEFA and DGESL in turn call three routines from the Basic Linear Algebra Subprograms, or BLAS [114]: IDAMAX, DSCAL, and DAXPY.

While the LINPACK routines DGEFA and DGESL involve two-dimensional array references, the BLAS refer to one-dimensional arrays. The LINPACK routines in general have been organized to access two-dimensional arrays by column. In DGEFA, for example, the call to DAXPY passes an address within the two-dimensional array A, which is then treated as a one-dimensional reference within DAXPY. Since the indexing is *down* a column of the two-dimensional array, the references to the one-dimensional array are sequential with unit stride. This is a performance enhancement over, say, addressing along the row of a two-dimensional array. Since Fortran dictates that two-dimensional arrays are stored by columns in memory, accesses to consecutive elements of a column lead to simple index calculations. References to consecutive elements differ by one word instead of by the leading dimension of the two-dimensional array.

4.5.3   Asymptotic Performance

We have run the LINPACK benchmark on a number of different machines and have collected data for various orders of matrices. This information can be used to give measures of $n_{1/2}$ and r_∞ for computing the matrix solution with the LINPACK programs, as follows.

We know that the algorithm uses $2/3n^3 + n^2 + O(n)$ operations for a matrix of order n. We also know that the time to complete the problem has the form $c_3n^3 + c_2n^2 + c_1n$. Hence, we can describe the rate of execution as the ratio of the operations performed to the time required to perform those operations, or

$$rate = \frac{2/3n^3 + n^2 + O(n)}{c_3n^3 + c_2n^2 + c_1n}. \tag{4.13}$$

We can compute the coefficients c_1, c_2, and c_3 through a least-squares fit of the data. From these coefficients we can then predict the asymptotic rate. At $n = \infty$ we have

$$r_\infty = \frac{2}{3c_3}. \tag{4.14}$$

By examining the point at which we reach half the asymptotic rate, we can determine $n_{1/2}$,

$$\frac{r_\infty}{2} = \frac{1}{3c_3} = \frac{2/3n^3 + \ldots}{c_3 n^3 + c_2 n^2 + \ldots}. \tag{4.15}$$

Solving for n, we determine that

$$n_{1/2} \approx c_2/c_3. \tag{4.16}$$

We compute the coefficients r_∞ and $n_{1/2}$ from the data. We then plot the data as well as the model and indicated values for r_∞ and $n_{1/2}$. In the following figures, the estimated asymptotic rate r_∞ is the horizontal dotted line, and the $n_{1/2}$ is the vertical dashed line. The circles represent the measured performance data for the LINPACK benchmark programs. The solid curve of Mflops is based on the model.

CRAY X-MP

The performance for the CRAY X-MP is an example of typical performance for a vectorizable algorithm over a range of problem sizes (see Figure 4.2). The performance for small problems is low; and as the size of the problem increases, the performance increases until it reaches its asymptotic rate.

Alliant

The performance of the Alliant FX/8 provides an interesting insight into the use of memory hierarchy. As the matrix size increases from order 1, we see the expected trend of increasing performance. When the size of the matrix exceeds 200, however, the performance decreases and eventually settles down and reaches an asymptotic rate less than the rate of a matrix of order 200. This result appears surprising at first glance, but is explained by looking at the memory hierarchy of the machine.

The machine has a 512-Kbyte cache. For a small-order matrix, the cache can accommodate the complete matrix and needs to load the matrix only once, where it will be reused from the cache. Since the cache is 512 Kbytes, it can accommodate 64,000 full-precision words or a matrix of order 252. Since other variables such as loop indices, scalar variables, and program code are also passed through the cache, the actual size of the matrix that can be fully contained is smaller than 252; and as we can see from the data, the peak is achieved around a matrix of order 210.

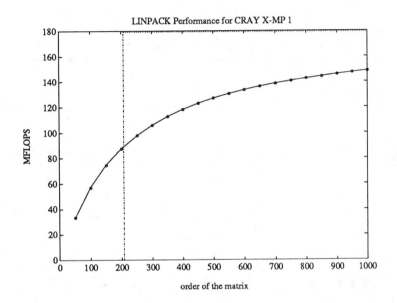

Figure 4.2: **LINPACK performance for the CRAY X-MP**

As the matrix becomes larger, it can no longer be fully contained in the cache, and parts must be reloaded from main memory. Thus, the rate of execution decreases as the matrix size increases past that of order 210. This trend continues until the asymptotic rate associated with moving the data fully from memory is reached (see Figure 4.3).

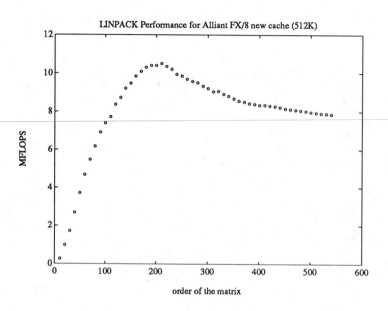

Figure 4.3: **LINPACK performance for Alliant FX/8**

Ardent Titan

The performance on the Ardent Titan provides another example, similar to that of the Alliant, of a problem with the use of the memory hierarchy. The rate of execution rises as the order of the matrix increases up to a matrix of order 550; then the performance decreases. The Ardent is a virtual-memory machine and uses a table look-aside buffer (TLB) to increase the performance of the paging system. The TLB contains the starting address of pages used in a program during execution. This configuration reduces the time required to resolve references to variables. At the same time, the TLB does not have enough locations to map all of physical memory. Thus, if a program references more bytes of data than the TLB holds, performance can be degraded by TLB misses. This situation occurs, for example, for a matrix of size 600 (see Figure 4.4).

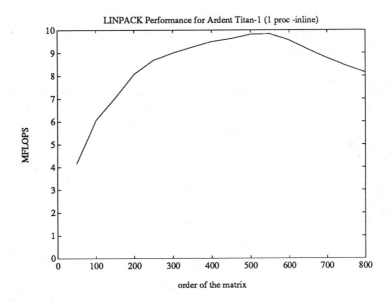

Figure 4.4: **LINPACK performance for the Ardent**

Summary of Machines

We have carried out the analysis for the asymptotic performance for running the LINPACK benchmark on a number of different machines. The results are summarized in Table 4.10.

On scalar machines, $n_{1/2}$ is effectively one, since there is no startup of vectors, and thus no penalty for short vectors. On machines that have a large value for $n_{1/2}$, we might expect to have difficulty in coming close to peak performance for general problems when the dominant operation is a simple vector operation such as a SAXPY.

Table 4.10: **Summary of $n_{1/2}$ and r_∞ for LINPACK Benchmark**

Machine	LINPACK $n = 100$, Mflops	$n_{1/2}$ Problem Size	r_∞ Mflops	Peak Mflops
Alliant FX/8 (512K cache)	6.9	160	7.3	94
Alliant FX/8 (128K cache)	5.9	34	7.6	94
Convex C-210	17	19	21	50
Convex C-220	21	82	42	100
Convex C-230		132	61	150
Convex C-240	27	145	74	200
CRAY X-MP/SE	53	87	99	210
CRAY X-MP/1	70	179	208	235
CRAY X-MP/4	178	179	456	940
CRAY Y-MP/1	144	100	208	333
CRAY Y-MP/8	275	434	1375	2667
CRAY-2S/1	44	305	138	488
CRAY-2S/4	101	297	362	1951
Cydrome Cydra 5	14	29	22	25
ETA-10P	27	484	167	167
ETA-10Q	34	484	211	210
ETA-10E	62	458	377	381
Fujitsu VP-400E	20	1522	316	1142
Fujitsu VP-400	20	1506	316	1142
Hitachi 810/20	17	2456	983	840
Hitachi 820/80	107	1693	1318	3000
Intel iPSC/2 (1 proc.)	.92	183	2.6	
NEC SX/2	43	1017	570	1300
Sequent Symmetry (1 proc.)	.21	24	.22	
Sun 3 w/FPA	.11	5	.13	
VAX 11/780 w/FPA UNIX	.12		.09	

The performance of the LINPACK benchmark is typical for applications where the basic operation is based on vector primitives. Some improvements may be possible by reorganizing or rearranging loop. However, in order to use the machine effectively and to obtain close to peak performance, such simple strategies as loop interchanging are not enough. In the next chapter, we discuss an approach that will achieve higher performance for the same number of floating-point operations.

Chapter 5

Building Blocks in Linear Algebra

One of the important aspects to utilizing a high-performance computer effectively is to avoid unnecessary memory references. In most computers, data flows from memory into and out of registers and from registers into and out of functional units, which perform the given instructions on the data. Performance of algorithms can be dominated by the amount of memory traffic, rather than the number of floating-point operations involved. The movement of data between memory and registers can be as costly as arithmetic operations on the data. This cost provides considerable motivation to restructure existing algorithms and to devise new algorithms that minimize data movement.

5.1 Basic Linear Algebra Subprograms

One way of achieving efficiency in the solution of linear algebra problems is through the use of the Basic Linear Algebra Subprograms. In 1973, Lawson et al. [114] described the advantages of adopting a set of basic routines for problems in linear algebra. The BLAS, as they are now commonly called, have been very successful and have been used in a wide range of software, including LINPACK and many of the algorithms published by the *ACM Transactions on Mathematical Software*. They are an aid to clarity, portability, modularity, and maintenance of software, and they have become a *de facto* standard for the elementary vector operations. The motivation for the BLAS is described in [114] and by Dodson and Lewis [27]. Here we review their purpose and their advantages. We also discuss two recent extensions to the BLAS. A complete list of routines, including calling sequence, and operations can be found in Appendix D.

The BLAS promote modularity by identifying frequently occurring operations of linear algebra and by specifying a standard interface to these operations. Efficiency may be achieved through optimization within the BLAS without altering the higher-level code that has referenced them. Obviously, it is important to identify and define a set of basic operations that are both rich enough to enable an expression of important high-level algorithms and simple enough to admit a very high level of optimization on a variety of computers. Such optimizations can often be achieved through modern compilation techniques, but hand coding in assembly language is also an option. Use of these optimized operations can yield dramatic reductions in computation time on some computers.

The BLAS also offer several other benefits:

- *Robustness* of linear algebra computations is enhanced by the BLAS, since they take into consideration algorithmic and implementation subtleties that are likely to be ignored in a typical application programming environment, such as treating overflow situations.

- Program *portability* is improved through standardization of computational kernels without giving up efficiency, since optimized versions of the BLAS can be used on those computers for which they exist, yet compatible standard Fortran is available for use elsewhere.

- Program *readability* is enhanced. The BLAS are a design tool; that is, they are a conceptual aid in coding, allowing one to visualize mathematical operations rather than the particular detailed coding required to implement the operations. By associating widely recognized mnemonic names with mathematical operations, the BLAS improve the self-documenting quality of code.

5.1.1   Level 1 BLAS

The original set of BLAS perform low-level operations such as dot-product and the adding of a multiple of one vector to another. We refer to these vector-vector operations as Level 1 BLAS. They provide a standard interface to a number of commonly used vector operations, and they instituted a useful naming convention that consists of a four- or five-letter mnemonic name preceded by one of the letters s, d, c, z to indicate precision type. Details are given in Appendix D.

The following types of basic vector operations are performed by the Level 1 BLAS:

$$y \leftarrow \alpha x + y \qquad\qquad dot \leftarrow x^T y$$
$$x \leftarrow \alpha x \qquad\qquad\qquad nrm2 \leftarrow ||x||_2$$
$$x \leftrightarrow y \qquad\qquad\qquad y \leftarrow x$$
$$asum \leftarrow ||re(x)||_1 + ||im(x)||_1 \quad amax \leftarrow 1^{st}k \ni |re(x_k)| + |im(x_k)| = ||x||_\infty$$

Generating plane rotations   Applying plane rotations

Typically these operations involve $O(n)$ floating-point operations and $O(n)$ data items moved (loaded or stored), where n is the length of the vectors.

The Level 1 BLAS permit efficient implementation on scalar machines, but the ratio of floating-point operations to data movement is too low to achieve effective use of most vector or parallel hardware. Even on a scalar machine, the cost of a subroutine call can be a performance issue when vector lengths are short and there is a limitation to the possible use of register allocations to improve the ratio of floating-point operations to data movement. The simplest and most central example is the computation of a matrix-vector product. This may be coded as a sequence of n SAXPY ($y \leftarrow \alpha x + y$) operations, but this obscures the fact that the result vector y could have been held in a vector register. Nevertheless, the clarity of code and the modularity contribute greatly to the ability to quickly construct correct, readable, and portable programs.

5.1.2   Level 2 BLAS

The principles behind the Level 1 BLAS are indeed sound and have served well in practice. However, with the advent of high-performance computers that use vector processing, it was soon recognized that additional BLAS would have to be created, since the Level 1 BLAS do not have sufficient *granularity* to admit optimizations such as reuse of data in registers and reduction in memory access. One needs to optimize at least at the level of matrix-vector operations in order to approach the potential efficiency of the machine; and the Level 1 BLAS inhibit this optimization, because they hide the matrix-vector nature of the operations from the compiler.

Thus, an additional set of BLAS, called the Level 2 BLAS, was designed for a small set of matrix-vector operations that occur frequently in the implementation of many of the most common algorithms in linear algebra [32]. The Level 2 BLAS involve $O(mn)$ scalar operations, where m and n are the dimensions of the matrix involved. The following three types of basic operation are performed by the Level 2 BLAS:

(a) Matrix-vector products of the form

$$y = \alpha A x + \beta y,$$
$$y = \alpha A^T x + \beta y, \text{ and}$$
$$y = \alpha \bar{A}^T x + \beta y$$

where α and β are scalars, x and y are vectors, and A is a matrix; and

$$x = T x,$$
$$x = T^T x, \text{ and}$$
$$x = \bar{T}^T x,$$

where x is a vector and T is an upper or lower triangular matrix.

(b) Rank-one and rank-two updates of the form

$A = \alpha x y^T + A,$
$A = \alpha x \bar{y}^T + A,$
$H = \alpha x \bar{x}^T + H,$ and
$H = \alpha x \bar{y}^T + \bar{\alpha} y \bar{x}^T + H,$

where H is a Hermitian matrix.

(c) Solution of triangular equations of the form

$x = T^{-1} x,$
$x = T^{-T} x,$ and
$x = \bar{T}^{-T} x,$

where T is a nonsingular upper or lower triangular matrix.

Where appropriate, the operations are applied to general, general band, Hermitian, Hermitian band, triangular, and triangular band matrices in both real and complex arithmetic and in single and double precision.

Many of the frequently used algorithms of numerical linear algebra can be coded so that the bulk of the computation is performed by calls to Level 2 BLAS routines; efficiency can then be obtained by using tailored implementations of the Level 2 BLAS routines. On vector-processing machines, one of the aims of such implementations is to keep the vector lengths as long as possible, and in most algorithms the results are computed one vector (row or column) at a time. In addition, on vector register machines, performance is increased by reusing the results of a vector register and not storing the vector back into memory.

Unfortunately, this approach to software construction is often not well suited to computers with a hierarchy of memory (such as global memory, cache or local memory, and vector registers) and true parallel-processing computers.

5.1.3   Level 3 BLAS

Initial experience with machines [43, 75] having a memory hierarchy to implement shared-memory parallel vector processing (such as the Alliant FX/80 computer) indicated that the Level 2 BLAS did not have a ratio of floating point to data movement that was high enough to make efficient reuse of data that resided in cache or local memory. For those architectures, it is often preferable to partition the matrix or matrices into blocks and to perform the computation by matrix-matrix operations on the blocks. By organizing the computation in this fashion, we provide for full reuse of data while the block is held in the cache or local memory. This approach avoids excessive movement of data to and from memory. In fact, it is often possible to obtain $O(n^3)$ floating-point operations while creating only $O(n^2)$ data movement. This phenomenon is often called the *surface-to-volume* effect for the ratio of operations to data movement. In addition, on architectures that allow parallel processing, parallelism can be exploited in two ways: (1) operations on distinct blocks may be performed in parallel; and (2) within the operations on each block, scalar or vector operations may be performed in parallel.

The Level 3 BLAS [31] are targeted at the matrix-matrix operations required for these purposes. The routines are derived in a fairly obvious manner from some of the Level 2 BLAS, by replacing the vectors x and y with matrices B and C. The advantage in keeping the design of the software as consistent as possible with that of the Level 2 BLAS is that it is easier for users to remember the calling sequences and parameter conventions.

In real arithmetic the operations for the Level 3 BLAS have the following forms:

(a) Matrix-matrix products:

$$C = \alpha AB + \beta C$$
$$C = \alpha A^T B + \beta C$$
$$C = \alpha AB^T + \beta C$$
$$C = \alpha A^T B^T + \beta C$$

These operations are more accurately described as matrix-matrix multiply-and-add operations; they include rank-k updates of a general matrix.

(b) Rank-k updates of a symmetric matrix:

$$C = \alpha AA^T + \beta C$$
$$C = \alpha A^T A + \beta C$$
$$C = \alpha A^T B + \alpha B^T A + \beta C$$
$$C = \alpha AB^T + \alpha BA^T + \beta C$$

(c) Multiplying a matrix by a triangular matrix:

$$B = \alpha TB$$
$$B = \alpha T^T B$$
$$B = \alpha BT$$
$$B = \alpha BT^T$$

(d) Solving triangular systems of equations with multiple right-hand sides:

$$B = \alpha T^{-1} B$$
$$B = \alpha T^{-T} B$$
$$B = \alpha BT^{-1}$$
$$B = \alpha BT^{-T}$$

Here α and β are scalars; A, B, and C are rectangular matrices (in some cases, square and symmetric); and T is an upper or lower triangular matrix (and nonsingular in (d)).

Analogous operations are in complex arithmetic: conjugate transposition is specified instead of simple transposition, and in (b) C is Hermitian and α and β are real.

The results of using the different levels of BLAS on the Alliant FX/8, IBM 3090/VF, and CRAY-2 are shown in Table 5.1.

Table 5.1: **Speed of the BLAS on Various Architectures** (all values are in Mflops)

	Alliant FX/8 (8 processors)	IBM 3090/VF (1 processor)	CRAY-2S (1 processor)
Peak Speed	94	108	488
LINPACK Benchmark	7.6	12	44
Level 1 BLAS $y = \alpha x + y$	14	26	121
Level 2 BLAS $y = \alpha A x + \beta y$	26	60	350
Level 3 BLAS $C = \alpha A B + \beta C$	43	80	437

The rates of execution quoted in Table 5.1 for the Level 1, 2, and 3 BLAS represent the asymptotic rates for operations implemented in assembler language. The performance of the LINPACK benchmark in Table 5.1 is limited to the performance of the Level 1 BLAS operations since the software is based on those operations.

Table 5.2 illustrates the advantage of the Level 3 BLAS through a comparison of the ratios of floating-point operations to data movement for three closely related operations from the Level 1, 2, and 3 BLAS. The second column counts the number of loads and stores required, while the third column counts the number of floating-point operations required to complete the operation. The fourth column reports the ratio of these two.

Table 5.2: **Advantage of the Level 3 BLAS**

BLAS	Loads and Stores	Floating-Point Operations	Ref:Ops (ratio) $n = m = k$
Level 1 SAXPY $y = y + \alpha x$	$3n$	$2n$	$3 : 2$
Level 2 SGEMV $y = \beta y + \alpha Ax$	$mn + n + 2m$	$2mn$	$1 : 2$
Level 3 SGEMM $C = \beta C + \alpha AB$	$2mn + mk + kn$	$2mnk$	$2 : n$

5.2 Levels of Parallelism

Today's advanced computers exhibit different levels of parallelism. In this section, we discuss these different levels and briefly outline how algorithms may be recast to achieve efficiency.

5.2.1 Vector Computers

Vector architectures exploit parallelism at the lowest level of computation. They require regular data structures (i.e., rectangular arrays) and large amounts of computation in order to be effective.

Three basic modes of execution are possible on vector computers: scalar, vector, and super-vector [36]. The basic difference between scalar and vector performance is the use of vector instructions. The difference between vector and super-vector performance hinges on avoiding unnecessary movement of data between vector registers and memory.

To provide a feeling for the difference in execution rates, we give in Table 5.3 the execution modes and rates for the CRAY-1.

Table 5.3: **Execution Modes and Rates on a CRAY-1**

Mode of Execution	Rate of Execution
Scalar	0 - 10 Mflops
Vector	11 - 50 Mflops
Super-Vector	51 - 160 Mflops

Most algorithms in linear algebra can be easily vectorized. However, to gain the most out of a machine like the CRAY-1, such vectorization is usually not enough. The CRAY-1 is limited in the sense that there is only one path between memory and the vector registers. This creates a bottleneck if a program loads a vector from memory, performs some arithmetic operations, and then stores the results. In order to achieve top performance, the scope of the vectorization must be expanded to facilitate chaining and minimize data movement, in addition to using vector operations. Recasting the algorithms in terms of matrix-vector operations makes it easy for a vectorizing compiler to achieve these goals.

Let us look more closely at how the recasting of an algorithm is performed. Many of the algorithms in linear algebra can be expressed in terms of a SAXPY operation. This results in three vector memory references for each two vector floating-point operations. If this operation constitutes the body of an inner loop that updates the same vector y many times, a considerable amount of unnecessary data movement will occur.

Usually, if a SAXPY occurs in an inner loop, the algorithm may be recast in terms of some matrix-vector operation, such as $y = y + M * x$, which is just a sequence of SAXPYs involving the columns of the matrix M and the corresponding components of the vector x (this is a GAXPY or generalized SAXPY operation). Thus it is relatively easy to recognize automatically that only the columns of M need be moved into registers while accumulating the result y in a vector register, avoiding two of the three memory references in the innermost loop. The rewritten algorithm also allows chaining to occur on vector machines. When the matrix-vector operations were hand-coded in assembler language, a factor of three increase in performance was obtained on the CRAY-1.

Since that time, a considerable improvement has been made in the ability of an optimizing compiler to vectorize and in some cases even to automatically parallelize the nested loops that occur in the BLAS. Usually, modern optimizing compilers are sophisticated enough to achieve near-optimal performance of the BLAS at all levels because of their regularity and simplicity. However, if the compiler is not successful, it is still quite reasonable to hand tune these operations, perhaps in assembly language, since there are so few of them and since they involve simple operations on regular data structures.

Higher-level codes may also be constructed from these modules. The resulting codes have achieved super-vector performance levels on a wide variety of vector architectures. Moreover, the same codes have also proved effective on parallel architectures. In this way portability has been achieved without suffering a serious degradation in performance.

5.2.2 Parallel Processors with Shared Memory

The next level of parallelism involves individual scalar processors executing serial instruction streams simultaneously on a shared-data structure. A typical example would be the simultaneous execution of a loop body for various values of the loop index. This is the capability provided by a parallel processor. With this increased functionality, however, comes a burden. This communication requires synchronization: if these independent processors are to work together on the same computation, they must be able to communicate partial results to each other. Such synchronization introduces overhead. It also requires new programming techniques that are not well understood at the moment. However, the simplicity of the basic operations within the Level 2 and Level 3 BLAS does allow the encapsulated exploitation of parallelism.

Typically, a parallel processor with globally shared memory must employ some sort of interconnection network so that all processors may access all of the shared memory. There must also be an arbitration mechanism within this memory access scheme to handle cases where two processors attempt to access the same memory location at the same time. These two requirements obviously have the effect of increasing the memory access time over that of a single processor accessing a dedicated memory of the same type.

Again, memory access and data movement dominate the computations in these machines. Achieving near-peak performance on such computers requires devising algorithms that minimize data movement and reuse data that has been moved from globally shared memory to local processor memory.

The effects of efficient data management on the performance of a parallel processor can be dramatic. For example, performance on the Denelcor HEP computer was increased by a factor of ten through efficient use of its (2K-word) register set [38]. A modular approach again aids in accomplishing this memory management. Moreover, modules provide a way to make effective use of the parallel processing capabilities in a manner that is transparent to the user of the software. Thus the user does not need to wrestle with the problems of synchronization in order to make effective use of the parallel processor.

5.2.3 Parallel-Vector Computers

The two types of parallelism we have just discussed are combined when vector rather than serial processors are used to construct a parallel computer. These machines are able to execute independent loop bodies that employ vector instructions. The most powerful computers today are of this type. They include the CRAY Y-MP line and the Alliant FX/80 and Convex C-2 high-performance mini-supercomputers. The problems with using such computers efficiently are, of course, more difficult than those encountered with each type individually. Synchronization overhead becomes more significant for a vector operation than a scalar operation, blocking loops to exploit outer-level parallelism may conflict with vector length, and so on.

5.2.4   Clusters of Parallel Vector Processors

A further complication is added when parallel vector machines are interconnected to achieve yet another level of parallelism. This is the case for the CEDAR architecture developed at the Center for Supercomputing Research and Development at the University of Illinois at Urbana [28]. Such a computer is intended to solve large applications problems that naturally split up into loosely coupled parts which may be solved efficiently on a cluster of parallel vector processors.

5.3   Basic Factorizations of Linear Algebra

In this section we develop algorithms for the solution of linear systems of equations and linear least-squares problems. Such problems are basic to all scientific and statistical calculations. Our intent here is to briefly introduce these basic notions. Thorough treatments may be found in [151, 84].

First, we consider the solution of the linear system

$$Ax = b, \tag{5.1}$$

where A is a real $n \times n$ matrix and x and b are both real vectors of length n. To construct a numerical algorithm, we rely on the fundamental result from linear algebra that implies that a permutation matrix P exists such that

$$PA = \mathrm{LU},$$

where L is a unit lower triangular matrix (ones on the diagonal) and U is upper triangular. The matrix U is nonsingular if and only if the original coefficient matrix A is nonsingular. Of course, this factorization facilitates the solution of equation (5.1) through the successive solution of the triangular systems

$$Ly = Pb, \; Ux = y.$$

The well-known numerical technique for constructing this factorization is called Gaussian elimination with partial pivoting. The reader is referred to [84] for a detailed discussion. As a point of reference, a brief development of this basic factorization will be given here. Variants of the factorization will provide a number of block algorithms.

5.3.1   Point Algorithm: Gaussian Elimination with Partial Pivoting

Let P_1 be a permutation matrix such that

$$P_1 A e_1 = \begin{pmatrix} \delta \\ c \end{pmatrix},$$

where $|\delta| \geq |c^T e_j|$ for all j (i.e., δ is the element of largest magnitude in the first column). If $\delta \neq 0$, put $l = c\delta^{-1}$; otherwise put $l = 0$. Then

$$P_1 A = \begin{pmatrix} \delta & u^T \\ c & \hat{A} \end{pmatrix} = \begin{pmatrix} 1 & \\ l & I \end{pmatrix} \begin{pmatrix} \delta & u^T \\ 0 & \hat{A} - lu^T \end{pmatrix}.$$

Suppose now that $\hat{L}\hat{U} = \hat{P}\left(\hat{A} - lu^T\right)$. Then

$$\begin{pmatrix} 1 & 0 \\ 0 & \hat{P} \end{pmatrix} P_1 A = \begin{pmatrix} 1 & 0 \\ \hat{P}l & \hat{L} \end{pmatrix} \begin{pmatrix} \delta & u^T \\ 0 & \hat{U} \end{pmatrix},$$

and

$$PA = LU,$$

where P, L, and U have the obvious meanings in the above factorization.

This discussion provides the basis for an inductive construction of the factorization $PA = LU$. However, a development that is closer to what is done in practice may be obtained by repeating the basic factorization step on the "reduced matrix" $\hat{A} - lu^T$ and continuing in this way until the final factorization has been achieved in the form

$$A = \left(P_1^T L_1 P_2^T L_2 ... P_{n-1}^T L_{n-1}\right) U,$$

with each P_i representing a permutation in the (i, k_i) positions where $k_i \geq i$. A number of algorithmic consequences are evident from this representation. One is that the permutation matrices may be represented by a single vector p with $p_i = k_i$. To compute the action $P_i b$, we simply interchange elements i and k_i. This representation leads to the following numerical procedure for the solution of (5.1).

> for $j = 1$ step 1 until n
> $b \leftarrow L_j^{-1} P_j b$; (1)
> end ;
> Solve $Ux = b$;

Since $L_j^{-1} = I - l_j e_j^T$, (1) amounts to a SAXPY operation.

5.3.2   Special Matrices

Often matrices that arise in scientific calculations will have special structure. There are good reasons for taking advantage of such structure when it exists. In particular, storage requirements can be reduced, the number of floating-point operations can be reduced, and more stable algorithms can be obtained.

An important class of such special matrices is symmetric matrices. A matrix is *symmetric* if $A = A^T$. A symmetric matrix A is *positive semidefinite* if $x^T A x \geq 0$ for all x, and it is *indefinite* if $(x^T A x)$ takes both positive and negative values for different vectors x. It is called *positive definite* if $x^T A x > 0$ for all $x \neq 0$. A well-known result is that a symmetric matrix A is positive definite if and only if

$$A = LDL^T,$$

where L is unit lower triangular and D is diagonal with positive diagonal elements.

This factorization results from a modification of the basic Gaussian elimination algorithm to take advantage of the fact that a matrix is symmetric and positive definite. This results in the Cholesky factorization.

Cholesky Factorization. The computational algorithm for Cholesky factorization can be developed as follows. Suppose that A is symmetric and positive definite. Then $\delta = e_1^T A e_1 > 0$. Thus, putting $l = a\delta^{-1}$ gives   where $a = (a_{12} \ a_{13} \cdots a_{1n})$, $\delta = a_{11}$

$$A = \begin{pmatrix} \delta & a^T \\ a & \hat{A} \end{pmatrix} = \begin{pmatrix} 1 & \\ l & I \end{pmatrix} \begin{pmatrix} \delta & \\ & \hat{A} - la^T \end{pmatrix} \begin{pmatrix} 1 & l^T \\ & I \end{pmatrix}.$$

Since $l = a\delta^{-1}$, the submatrix $\hat{A} - la^T = \hat{A} - a(\delta^{-1})a^T$ is also symmetric and is easily shown to be positive definite as well. Therefore, the factorization steps may be continued (without pivoting) until the desired factorization is obtained. Details about numerical stability and variants are given in [132]. $\hookrightarrow A = LDL^T$, then $C = L\sqrt{D}$, $A = CC^T$

Symmetric Indefinite Factorization. When A is symmetric but indefinite, pivoting must be done to factor the matrix. The example

$$\begin{pmatrix} 0 & 1 \\ 1 & 0 \end{pmatrix}$$

shows that there is no general factorization of the form

$$A = LDL^T,$$

where D is diagonal and L is lower triangular when A is indefinite. There is, however, a closely related factorization

$$PAP^T = LDL^T,$$

where P is a permutation matrix, L is a unit lower triangular matrix, and D is a block diagonal matrix with one-by-one and two-by-two diagonal blocks.

Bunch and Kaufman [19] have devised a clever partial pivoting strategy to construct such a factorization. For a single step, the pivot selection is completely determined by examining two columns of the matrix A, with the selection of the second column depending on the location of the maximum element of the first column of A. Let j be the index of the entry in column 1 of largest magnitude. The pivot matrix is the identity if no pivot is required; an interchange between row and column 1 with row and column j if entry α_{jj} is to become a 1×1 pivot; and an interchange between row and column 2 with row and column j if the submatrix

$$\begin{pmatrix} \alpha_{11} & \alpha_{j1} \\ \alpha_{j1} & \alpha_{jj} \end{pmatrix}$$

is to become a 2×2 pivot. The pivot strategy devised by Bunch and Kaufman [19] is described in the following pseudo-code. In this code $\theta = (1 + (17)^{1/2})/8$ is chosen to optimally control element growth in the pivoting strategy:

$j = index\ of\ max_i(|\alpha_{i1}|);$
$\mu = max_j(\ |\alpha_{jl}|);$
if ($|\alpha_{11}| \geq \theta|\alpha_{j1}|$ or $|\alpha_{11}|\mu \geq \theta|\alpha_{j1}|^2$) then
    $j = 1;$ (1×1 pivot; with no interchange);
else
    if ($|\alpha_{jj}| \geq \theta\mu$) then
        (1×1 pivot; interchange 1 with j);
    else
        (2×2 pivot; interchange 2 with j);
    endif;
endif;
end;

This scheme is actually a slight modification of the original Bunch-Kaufman strategy. It has the same error analysis and has a particularly nice feature for positive definite matrices. We emphasize two important points. First, the column $j > 1$ of A—which must be examined in order to determine the necessary pivot—is determined solely by the index of the element in the first column that is of largest magnitude. Second, if the matrix A is positive definite, then *no pivoting will be done*, and the factorization reduces to the Cholesky factorization. To see this, note the following. If A is positive definite, then the submatrix

$$\begin{pmatrix} \alpha_{11} & \alpha_{j1} \\ \alpha_{j1} & \alpha_{jj} \end{pmatrix}$$

must be positive definite. Thus,

$$|\alpha_{11}|\mu = \alpha_{11}\mu \geq \alpha_{11}\alpha_{22} > \alpha_{j1}^2 > \theta\alpha_{j1}^2,$$

which indicates that α_{11} will be a 1×1 pivot and no interchange will be made according to the pivot strategy. Further detail concerning this variant is available in [150].

Now, assume that the pivot selection for a single step has been made. If a 1×1 pivot has been indicated, then a symmetric permutation is applied to bring the selected diagonal element to the (1,1) position, and the factorization step is exactly as described above for the Cholesky factorization. If a 2×2 pivot is required, then a symmetric permutation must be applied to bring the element of largest magnitude in the first column into the (2,1) position. Thus

$$P_1 A P_1^T = \begin{pmatrix} D & C \\ C^T & \hat{A} \end{pmatrix} = \begin{pmatrix} I & \\ CD^{-1} & I \end{pmatrix} \begin{pmatrix} D & \\ & \hat{A} - CD^{-1}C^T \end{pmatrix} \begin{pmatrix} 1 & D^{-1}C^T \\ & I \end{pmatrix}.$$

In this factorization step, let $D = \begin{pmatrix} \delta_1 & \beta \\ \beta & \delta_2 \end{pmatrix}$ and $C = (c_1, c_2)$. When a 2×2 pivot has been selected, it follows that $|\delta_1\delta_2| < \theta^2\beta^2$, so that $det(D) < (\theta^2 - 1)\beta^2 < 0$ and D is a 2×2 indefinite matrix.

Note that with this strategy every 2×2 pivot matrix D has a positive and a negative eigenvalue. Thus one can read off the inertia (the number of positive, negative, and zero eigenvalues) of the factorized matrix by counting the number of positive 1×1 pivots plus the number of 2×2 pivots to get the number of positive eigenvalues, and the number of negative 1×1 pivots plus the number of 2×2 pivots to get the number of negative eigenvalues.

As with the other factorizations, this one may be continued to completion by repeating the basic step on the reduced submatrix

$$\hat{\hat{A}} = \hat{A} - CD^{-1}C^T.$$

Exploitation of symmetry reduces the storage requirement and also the computational cost by half.

An alternative to this factorization is given by Aasen in [1]. A blocked version of that alternative algorithm is presented in [150].

5.4 Blocked Algorithms: Matrix-Vector and Matrix-Matrix Versions

The basic algorithms just described are the core subroutines in LINPACK for solving linear systems. The original version of LINPACK was designed to use vector operations. However, the algorithms did not perform near the expected level on the powerful vector machines that were developed just as the package had been completed. The key to achieving high performance on these advanced architectures has been to recast the algorithms in terms of matrix-vector and matrix-matrix operations to permit reuse of data.

To derive these variants, we examine the implications of a block factorization in progress. One can construct the factorization by analyzing the way in which the various pieces of the factorization interact. Let us consider the decomposition of the matrix A into its LU factorization with the matrix partitioned in the following way. Let us suppose that we have factored A as $A = LU$. We write the factors in block form and observe the consequences.

$$
\begin{pmatrix}
A_{11} & A_{12} & A_{13} \\
A_{21} & A_{22} & A_{23} \\
A_{31} & A_{32} & A_{33}
\end{pmatrix}
=
\begin{pmatrix}
L_{11} & & \\
L_{21} & L_{22} & \\
L_{31} & L_{32} & L_{33}
\end{pmatrix}
\begin{pmatrix}
U_{11} & U_{12} & U_{13} \\
& U_{22} & U_{23} \\
& & U_{33}
\end{pmatrix}
$$

Multiplying L and U together and equating terms with A, we have

$$
\begin{aligned}
A_{11} &= L_{11}U_{11}, & A_{12} &= L_{11}U_{12}, & A_{13} &= L_{11}U_{13}, \\
A_{12} &= L_{21}U_{11}, & A_{22} &= L_{21}U_{12} + L_{22}U_{22}, & A_{23} &= L_{21}U_{13} + L_{22}U_{23}, \\
A_{31} &= L_{31}U_{11}, & A_{32} &= L_{31}U_{12} + L_{32}U_{22}, & A_{33} &= L_{31}U_{13} + L_{32}U_{23} + L_{33}U_{33}.
\end{aligned}
$$

With these simple relationships we can develop variants by postponing the formation of certain components and also by manipulating the order in which they are formed. A crucial factor for performance is the choice of the *blocksize, k* (i.e., the column width) of the second block column. A blocksize of 1 will produce matrix-vector algorithms, while a blocksize of $k > 1$ will produce matrix-matrix algorithms. Machine-dependent parameters such as cache size, number of vector registers, and memory bandwidth will dictate the best choice for the blocksize.

Three natural variants occur: right-looking, left-looking, and Crout (see Figure 5.1). The terms right and left refer to the regions of data access, and Crout represents a hybrid of the left- and right-looking version.

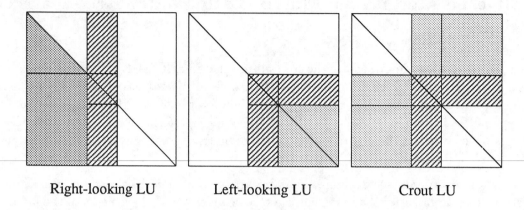

Right-looking LU   Left-looking LU   Crout LU

Figure 5.1: **Memory access patterns for variants of LU decomposition**

The shaded parts indicate the matrix elements accessed in forming a block row or column, and the darker shading indicates the block row or column being computed. The left-looking variant computes one block column at a time, using previously computed columns. The right-looking variant (the familiar recursive algorithm) computes a block row and column at each step and uses them to update the trailing submatrix. The Crout variant is a hybrid algorithm in which a block row and column are computed at each step, using previously computed rows and previously computed columns. These variants have been called the *i,j,k variants* owing to the arrangement of loops in the algorithm. For a more complete discussion of the different variants, see [36, 132].

Each of these can be modified to produce a variant of the Cholesky factorization when the matrix is known to be symmetric and positive definite. These modifications are straightforward and will not be presented. The symmetric indefinite case does provide some difficulty, however, and this will be developed separately.

Let us now develop these block variants of Gaussian elimination with partial pivoting.

5.4.1   Right-Looking Algorithm

The right-looking variant is really a blocked analogue of the point algorithm developed previously. However, we shall derive it from a slightly different viewpoint in order to set the stage for deriving the other two variants.

Suppose for the moment that a partial factorization of A has been obtained so that

$$PA = \begin{pmatrix} L_{11} & & \\ L_{21} & I & \\ L_{31} & 0 & I \end{pmatrix} \begin{pmatrix} U_{11} & U_{12} & U_{13} \\ & \hat{A}_{22} & \hat{A}_{23} \\ & \hat{A}_{32} & \hat{A}_{33} \end{pmatrix}.$$

Now, to advance the factorization, factor

$$P_2 \begin{pmatrix} \hat{A}_{22} \\ \hat{A}_{32} \end{pmatrix} = \begin{pmatrix} L_{22} \\ L_{32} \end{pmatrix} U_{22}$$

then

$$\begin{pmatrix} \hat{A}_{23} \\ \hat{A}_{33} \end{pmatrix} \leftarrow P_2 \begin{pmatrix} \hat{A}_{23} \\ \hat{A}_{33} \end{pmatrix}, \quad \begin{pmatrix} L_{21} \\ L_{31} \end{pmatrix} \leftarrow P_2 \begin{pmatrix} L_{21} \\ L_{31} \end{pmatrix}$$

and solve

$$U_{23} = L_{22}^{-1} \hat{A}_{23} \text{ (solution of triangular systems).}$$

Finally,

$$\hat{A}_{33} \leftarrow \hat{A}_{33} - L_{32} U_{23} \text{ (matrix-matrix product).}$$

At this point

$$P_2 P A = \begin{pmatrix} L_{11} & & \\ L_{21} & L_{22} & \\ L_{31} & L_{32} & I \end{pmatrix} \begin{pmatrix} U_{11} & U_{12} & U_{13} \\ & U_{22} & U_{23} \\ & & \hat{A}_{33} \end{pmatrix}.$$

Now, one simply needs to relabel the blocks to advance to the next block step. Note how the pivoting must be handled if the block size is greater than 1. It is no longer convenient to associate a pivot operation with each column. At best, one must maintain a block of pivot information associated with each block column reduction step.

The main advantage of this arrangement is during the formation of

$$\hat{A}_{33} \leftarrow \hat{A}_{33} - L_{32} U_{23},$$

which is a matrix-matrix operation if the block size is greater than 1. However, if the block size is equal to 1, then the Level 2 operation is to perform an outer product—generally the least efficient of the Level 2 BLAS since it updates the whole submatrix.

5.4.2 Left-Looking Algorithm

As we shall see, from the standpoint of data access, the left-looking variant is the best of the three. To begin, we assume that

$$
PA = \begin{pmatrix} L_{11} & & \\ L_{21} & I & \\ L_{31} & 0 & I \end{pmatrix} \begin{pmatrix} U_{11} & A_{12} & A_{13} \\ 0 & A_{22} & A_{23} \\ 0 & A_{32} & A_{33} \end{pmatrix}.
$$

To advance the factorization, we solve first

$$
U_{12} = L_{11}^{-1} A_{12} \text{ (solution of triangular systems)}
$$

and then

$$
\begin{pmatrix} \hat{A}_{22} \\ \hat{A}_{32} \end{pmatrix} \leftarrow \begin{pmatrix} A_{22} \\ A_{32} \end{pmatrix} - \begin{pmatrix} L_{21} \\ L_{31} \end{pmatrix} U_{12} \text{ (matrix-matrix product).}
$$

Finally, we factor

$$
P_2 \begin{pmatrix} \hat{A}_{22} \\ \hat{A}_{32} \end{pmatrix} = \begin{pmatrix} L_{22} \\ L_{32} \end{pmatrix} U_{22}
$$

and replace

$$
\begin{pmatrix} A_{23} \\ A_{33} \end{pmatrix} \leftarrow P_2 \begin{pmatrix} A_{23} \\ A_{33} \end{pmatrix} \text{ and } \begin{pmatrix} L_{21} \\ L_{31} \end{pmatrix} \leftarrow P_2 \begin{pmatrix} L_{21} \\ L_{31} \end{pmatrix}.
$$

At this point,

$$
P_2 PA = \begin{pmatrix} L_{11} & & \\ L_{21} & L_{22} & \\ L_{31} & L_{32} & I \end{pmatrix} \begin{pmatrix} U_{11} & U_{21} & A_{13} \\ 0 & U_{22} & A_{23} \\ 0 & 0 & A_{33} \end{pmatrix}.
$$

Observe that data accesses all occur to the left of the block column being updated. Moreover, the only write access occurs within this block column. Matrix elements to the right are referenced only for pivoting purposes, and even this procedure may be postponed until needed with a simple rearrangement of the above operations. Also, note that the original array A may be used to store the factorization, since the L is unit lower triangular and U is upper triangular. Of course, in this and all of the other versions of triangular factorization, the additional zeros and ones appearing in the representation do not need to be stored explicitly.

5.4.3   Crout Algorithm

The Crout variant is best suited for vector machines with enough memory bandwidth to support the maximum computational rate of the vector units. Its advantage accrues from avoiding the solution of triangular systems and also from requiring fewer memory references than the right-looking algorithm. Here, it is assumed that

$$
PA = \begin{pmatrix} L_{11} & & \\ 0 & I & \\ 0 & 0 & I \end{pmatrix} \begin{pmatrix} I & U_{12} & U_{13} \\ L_{21} & A_{22} & A_{23} \\ L_{31} & A_{32} & A_{33} \end{pmatrix} \begin{pmatrix} U_{11} & 0 & 0 \\ & I & 0 \\ & & I \end{pmatrix}.
$$

To advance this factorization, we solve

$$
(A_{22}, A_{23}) \leftarrow (A_{22}, A_{23}) - L_{21}(U_{12}, U_{13}) \text{ (matrix-matrix product)},
$$

then

$$
A_{32} \leftarrow A_{32} - L_{31}U_{12} \text{ (matrix-matrix product)},
$$

and then factor

$$
P_2 \begin{pmatrix} A_{22} \\ A_{32} \end{pmatrix} = \begin{pmatrix} L_{22} \\ L_{32} \end{pmatrix} U_{22}.
$$

Now we replace

$$
\begin{pmatrix} A_{23} \\ A_{33} \end{pmatrix} \leftarrow P_2 \begin{pmatrix} A_{23} \\ A_{33} \end{pmatrix}, \quad \begin{pmatrix} L_{21} \\ L_{31} \end{pmatrix} \leftarrow P_2 \begin{pmatrix} L_{21} \\ L_{31} \end{pmatrix}
$$

and put

$$U_{23} = L_{21}^{-1} A_{23}.$$

At this point,

$$P_2 P A = \begin{pmatrix} L_{11} & & \\ L_{21} & L_{22} & \\ 0 & 0 & I \end{pmatrix} \begin{pmatrix} I & & U_{13} \\ 0 & I & U_{23} \\ L_{31} & L_{32} & A_{33} \end{pmatrix} \begin{pmatrix} U_{11} & U_{12} & 0 \\ & U_{22} & 0 \\ & & I \end{pmatrix}.$$

Again, a repartitioning and relabeling will allow the factorization to advance to completion.

With this presentation, the storage scheme may be somewhat obscure. Note that the lower triangular factor L_{11} and the upper triangular factor U_{11} may occupy the upper left-hand corner of the original $n \times n$ array. Thus, the original array A may be overwritten with this factorization as it is computed.

5.4.4   Typical Performance of Blocked LU Decomposition

Most of the computational work for the LU variants is contained in three routines: the matrix-matrix multiply, the triangular solve with multiple right-hand sides, and the unblocked LU factorization for operations within a block column. Table 5.4 shows the distribution of work among these three routines and the average performance rates on one processor of a CRAY-2 for a sample matrix of order 500 and a blocksize of 64. In these examples, each variant calls its own unblocked variant, and the pivoting operation uses about 2% of the total time. The average speed of the matrix-matrix multiply on the CRAY-2 is over 400 Mflops for all three variants, but the average speed of the triangular solve depends on the size of the triangular matrices. For the left-looking variant, the triangular matrices at each step range in size from k to $n - k$, where k is the blocksize and n the order of the matrix, and the average performance is 268 Mflops, while for the right-looking and Crout variants, the triangular matrices are always of order k and the average speed is only 105 Mflops. Clearly the average performance of the Level 3 BLAS routines in a blocked routine is as important as the percentage of Level 3 BLAS work.

Table 5.4: **Breakdown of Operations and Times for LU Variants for** $n = 500$, $k = 64$
(CRAY-2S, 1 processor)

Variant	Routine	% Operations	% Time	Avg. Mflops
Left-looking	SGEMM } BLAS 3	49	32	438
	STRSM }	41	45	268
	unblocked LU	10	20	146
Right-looking	SGEMM	82	56	414
	STRSM	8	23	105
	unblocked LU	10	19	151
Crout	SGEMM	82	57	438
	STRSM	8	24	105
	unblocked LU	10	16	189

Despite the differences in the performance rates of their components, the block variants of the LU factorization tend to show similar overall performance, with a slight advantage to the right-looking and Crout variants because more of the operations are in SGEMM. This points out the importance of having a good implementation for all the BLAS.

5.4.5 Blocked Symmetric Indefinite Factorization

To derive a blocked version of this algorithm, we adopt the right-looking or rank-k update approach. The pivoting requirement apparently prevents the development of an effective left-looking algorithm.

To this end, suppose that $k - 1$ columns of the matrix L have been computed and stored in an $n \times (k - 1)$ array L. Then

$$PAP^T = LDL^T + \begin{pmatrix} 0 & 0 \\ 0 & \tilde{A} \end{pmatrix}$$

which may be rearranged to give

$$PAP^T - LDL^T = \begin{pmatrix} 0 & 0 \\ 0 & \tilde{A} \end{pmatrix}.$$

Now, to advance the factorization one step, we need only know the first and the jth column of $\tilde{A}$ where j is the index of the element of maximum magnitude in the first column. These two columns can be revealed without knowing $\tilde{A}$ explicitly, by noting that the ith column is given by

$$\tilde{A}e_i = \left[PAP^T - LDL^T \right] e_{k+i-1}.$$

We compute P_2 as shown above in the point algorithm to produce

$$P_2 \tilde{A} P_2^T = \begin{pmatrix} I & 0 \\ L_{22} & I \end{pmatrix} \begin{pmatrix} D_2 & 0 \\ 0 & \tilde{\tilde{A}} \end{pmatrix} \begin{pmatrix} I & L_{22}^T \\ 0 & I \end{pmatrix}.$$

Now,

$$P \leftarrow P \begin{pmatrix} I & 0 \\ 0 & P_2 \end{pmatrix}, \quad D \leftarrow \begin{pmatrix} D & 0 \\ 0 & D_2 \end{pmatrix}, \quad \text{and } L \leftarrow \begin{pmatrix} L_{11} & 0 \\ L_{21} & L_{22} \end{pmatrix}.$$

At this point the decomposition (5.1) has been updated with L becoming an $n \times k$ or an $n \times (k+1)$ matrix and D becoming a $k \times k$ or a $(k+1) \times (k+1)$ matrix according to the pivot requirements. Once the number of columns of L has become as large as desired, we let

$$PAP^T = \begin{pmatrix} A_{11} & A_{12}^T \\ A_{21} & A_{22} \end{pmatrix},$$

where the first block column has the same dimensions as L. Now the transformations may be explicitly applied with the trailing submatrix A_{22} being overwritten with

$$A_{22} \leftarrow A_{22} - L_{21} D L_{21}^T,$$

which can be computed by using a matrix-matrix product. Of course, symmetry will be taken advantage of here to reduce the operation count by half.

As shown above, it is possible to develop the left-looking algorithm in the case of LU decomposition, a procedure that is advantageous because it references only the matrix within a single block column for reads and writes and references only the matrix from the left for reads. The matrix entries to the right of the *front* are not referenced. An analogous version is apparently not possible for the symmetric indefinite factorization.

Let us demonstrate by trying to generate this version in the same manner as one would generate "left-looking" LU decomposition. The derivation requires a different approach to be taken. Suppose that

$$PAP^T = \begin{pmatrix} L_{11} & & \\ L_{21} & I & \\ L_{31} & 0 & I \end{pmatrix} \begin{pmatrix} DL_{11}^T & A_{12} & A_{13} \\ 0 & A_{22} & A_{23} \\ 0 & A_{32} & A_{33} \end{pmatrix}, \tag{5.2}$$

where the block

$$\begin{pmatrix} A_{12} \\ A_{22} \\ A_{32} \end{pmatrix}$$

is to be factored and used to update the existing factorization. Suppose for the moment that no additional pivoting is necessary. Then the steps required to update the factorization are as follows:

1. Note $L_{11}DL_{21}^T = A_{12}$ by symmetry.

2. Overwrite $A_{22} \leftarrow A_{22} - L_{21}DL_{21}^T$.

3. Overwrite $A_{32} \leftarrow A_{32} - L_{31}DL_{21}^T$.

4. Factor $A_{22} = L_{22}D_2L_{22}^T$.

5. Overwrite $L_{32} = A_{32} \leftarrow A_{32}D_2^{-1}L_{22}^{-T}$.

Now update (5.2) to obtain

$$PAP^T = \begin{pmatrix} L_{11} & 0 & 0 \\ L_{21} & L_{22} & 0 \\ L_{31} & L_{32} & I \end{pmatrix} \begin{pmatrix} DL^T{}_{11} & DL_{21}^T & A_{13} \\ 0 & D_2L_{22}^T & A_{23} \\ 0 & 0 & A_{33} \end{pmatrix}.$$

This advancement of the factorization by blocks may be continued by repartitioning and repeating these steps to adjoin the next block of columns.

Unfortunately, because of the required look-ahead, pivoting must be done one step at a time in order to fill out the block. The same look-ahead strategy described for the "right-looking" algorithm could be employed to construct the block one step at a time. However, the advantage of the apparent matrix-matrix products at steps 2 and 3 above is lost, because the pivoting may incorporate entries from the blocks A_{13}, A_{23}, and A_{33}. Thus, when pivoting has been incorporated, one is limited to matrix-vector operations. Hence, pivoting prevents the development of a "left-looking" Level 3 version of the symmetric indefinite factorization.

5.4.6 Typical Performance of Blocked Symmetric Indefinite Factorization

The performance for the symmetric indefinite factorization routine is along the same lines as LU decomposition. In Table 5.5 we show the increase in performance of the blocked version over the unblocked version for various orders of matrices. As can be seen, the performance of the blocked version is superior to the matrix-vector version.

Table 5.5: **Performance in Mflops of Symmetric Indefinite Factorization** $k = 64$ **(CRAY-2/S, 1 processor)**

n	Unblocked	Blocked
100	44	48
200	88	109
300	115	168
400	132	198
500	140	238
700	158	265
1000	171	314

5.5 Linear Least Squares

In addition to solving square linear systems, we are also concerned with overdetermined systems. That is, we wish to "solve" systems

$$Ax = b, \tag{5.3}$$

where A is a real $m \times n$ matrix with $m > n$, x is a vector of length n, and b is a vector of length m. Of course, the system (5.3) is consistent (i.e., has a solution) if and only if b is in the range of A. Therefore, we often seek a relaxed solution to a related problem:

$$\min_x ||Ax - b||. \tag{5.4}$$

This problem is called the linear least-squares problem when the 2-norm is used. We already have enough tools to solve this problem, since the quadratic function

$$||Ax - b||^2 = x^T(A^TA)x - 2b^TAx + b^Tb \tag{5.5}$$

has a unique minimum at

$$\hat{x} = (A^TA)^{-1}A^Tb, \tag{5.6}$$

which may be verified by substitution or by taking the gradient of (5.5) with respect to the components of x and setting it to zero. It is easily verified that $\hat{x}$ is the unique minimizer of (5.4) if and only if the symmetric matrix $A^T A$ is positive definite. Thus we can solve (5.4) by forming $A^T A$ and computing its Cholesky factorization.

5.5.1 Householder Method

The *normal equation* approach that we have just outlined is often used and is usually successful. The approach does, however, have drawbacks stemming from the potential loss of information through the explicit formation of $A^T A$ in finite-precision arithmetic and through the potential squaring of the *condition number* of the problem. When A has linearly independent columns, the condition number of a linear system or a linear least-squares problem with a coefficient matrix A is defined to be $\|A\|\,\|A^I\|$, where $A^I = A^{-1}$ when A is square and $A^I = (A^T A)^{-1} A^T$ when A has more rows than columns. The size of the condition number is a measure of the sensitivity of the solution to perturbations in the data (i.e., in the matrix A or in the right-hand side b). Further details may be found in [151].

An alternative has therefore been developed that is only slightly more expensive and that avoids these problems. The method is QR factorization with Householder transformations. Given a real $m \times n$ matrix A, the routine must produce an $m \times m$ orthogonal matrix Q and an $n \times n$ upper triangular matrix R such that

$$A = Q \begin{pmatrix} R \\ 0 \end{pmatrix}.$$

Householder's method involves constructing a sequence of transformations of the form

$$I - \alpha u u^T, \text{ where } \alpha u^T u = 2. \tag{5.7}$$

The vector u is constructed to transform the first column of the given matrix into a multiple of the first coordinate vector e_1. At the kth stage of the algorithm, one has

$$Q_{k-1}^T A = \begin{pmatrix} R_{k-1} & S_{k-1} \\ 0 & A_{k-1} \end{pmatrix},$$

and u_k is constructed such that

$$(I - \alpha_k u_k u_k^T) A_{k-1} = \begin{pmatrix} \rho_k & s_k^T \\ 0 & A_k \end{pmatrix}. \tag{5.8}$$

The factorization is then updated to the form

$$Q_k^T A = \begin{pmatrix} R_k & S_k \\ 0 & A_k \end{pmatrix}$$

with

$$Q_k = Q_k \begin{pmatrix} I & 0 \\ 0 & I - \alpha_k u_k u_k^T \end{pmatrix}.$$

However, the matrix Q_k is generally not explicitly formed, since it is available in product form if the vectors u are simply recorded in place of the columns they have been used to annihilate. This is the basic algorithm used in LINPACK [30] for computing the QR factorization of a matrix. The algorithm may be coded in terms of two of the Level 2 BLAS. To see this, just note that the operation of applying a transformation shown on the left-hand side of (5.8) may be broken into two steps:

$$z^T = u^T A \text{ (vector } \times \text{ matrix)} \tag{5.9}$$

and

$$\hat{A} = A - \alpha u z^T \text{ (rank-one modification)}.$$

5.5.2   Blocked Householder Method

One can develop a blocked version of the Householder method. The essential idea is to accumulate several of the Householder transformations into a single block transformation and then apply the result to annihilate several columns of the matrix at once. Both a left-looking and right-looking version of this algorithm can be constructed. We present the right-looking version.

To see how to construct such an aggregated transformation, note that (in general) it is possible to show that the product of k Householder transformations may be written as

$$\Pi(I - \alpha_j u_j u_j^T) = I - YTY^T,$$

where $Y = (u_1, u_2, ..., u_k)$ and T is an upper triangular matrix. This fact is easily verified by induction. If one has $V = I - YTY^T$, then

$$V(I - \alpha u u^T) = I - (Y, u) \begin{pmatrix} T & h \\ 0 & \alpha \end{pmatrix} (Y, u), \tag{5.10}$$

where $h = \alpha T W^T u$. One may also represent and compute this blocked transformation in a *WY representation* where $W = YT$. This latter representation, developed by Bischof and Van Loan [17], was later modified by Schreiber and Van Loan [143] into the *compact WY representation* presented above. The compact representation has the advantage of requiring less storage. The formation of

$$Vw = w - YT(Y^T w)$$

is rich in matrix-vector operations. One may accumulate the desired number of Householder trans-formations and then apply this block orthogonal transformation to obtain

$$(I - YTY^T)A = A - YT(Y^TA) = \begin{pmatrix} R_k & S_k^T \\ 0 & A_k \end{pmatrix}. \qquad (5.11)$$

Of course, only the last $n - k$ columns of A would be involved in this computation, with the first k columns computed as in the point algorithm. An interesting, but minor point in practice is that the block algorithm will require more floating-point operations than the point algorithm since it requires $O(mk^2)$ additional operations to accumulate the block factorization and $k^2/2$ extra storage locations for the matrix T. Typically, however, for all but the final stages of the factorization, m is much greater than k. Therefore, the contribution of these terms is of low order in comparison to the primary factorization costs. This situation does introduce some partitioning and load-balance problems in terms of parallel processing. Typically, the later stages of any triangular factorization will create a serial bottleneck, given the low order of the matrix that remains to be factored. Thus, the vector lengths will be short and the matrix sizes small for the final few factorization steps. This situation brings up the question of when to discontinue the block algorithm in favor of the point algorithm. It also brings up the problem of varying the block size throughout the factorization. These issues are discussed in some detail in [17].

5.5.3 Typical Performance of the Blocked Householder Factorization

Table 5.6 shows the performance in Mflops of four variants of the QR decomposition on one pro-cessor of a CRAY-2. The two block variants are SQRBR, a block right-looking algorithm in which a block Householder matrix is computed and immediately applied to the rest of the matrix as a rank-k update, and SQRBL, a block left-looking variant in which the previous updates are first applied to the current block column before the next block Householder matrix is computed. SQR2 is the unblocked Level 2 BLAS variant, and SQRDC is the Level 1 BLAS variant from LINPACK. We see that the blocked variants surpass the unblocked variant in performance only for matrices of order greater than 200. This is a result of a larger operation count in the blocked version.

Table 5.6: **Performance in Mflops of QR Variants, $k = 64$ (CRAY-2/S, 1 processor)**

	Matrix size $m = n$				
QR variant	100	200	300	400	500
LINPACK – SQRDC (vector version)	24	41	55	66	76
SQR2 (matrix-vector version)	144	215	242	251	255
SQRBR (matrix version - right looking)	106	209	269	306	328
SQRBL (matrix version - left looking)	102	198	258	293	316

5.6   Organization of the Modules

In this section we discuss the organization of code within the modules that make up the BLAS. A complete description would require far more detail than is appropriate for this book. Therefore, we shall limit ourselves to a discussion of matrix-vector and matrix-matrix products. Our objective is to give a generic idea of the considerations required for enhanced performance.

5.6.1   Matrix-Vector Product

We begin with some of the considerations that are important when coding the matrix-vector product module. The other modules require similar techniques. For vector machines such as the Cray series, the matrix-vector operation should be coded in the form

$$y(1:m) = y(1:m) + M(1:m,j)x(j) \text{ for } j = 1, 2, ..., n. \tag{5.12}$$

The $1:m$ in the first entry implies that this is a column operation. The intent here is to reserve a vector register for the result while the columns of M are successively read into vector registers, multiplied by the corresponding component of x, and then added to the result register in place. In terms of ratios of data movement to floating-point operations, this arrangement is most favorable. It involves one vector move for two vector floating-point operations. The advantage of using the matrix-vector operation is clear when we compare the result to the three vector moves required to get the same two floating-point operations when a sequence of SAXPY operations is used. This operation is sometimes referred to as a GAXPY operation [36].

This arrangement is perhaps inappropriate for a parallel machine, however, because one would have to synchronize the access to y by each of the processes, and such synchronization would cause busy waiting to occur. One might do better to partition the vector y and the rows of the matrix M into blocks:

$$\begin{pmatrix} y_1 \\ y_2 \\ \cdot \\ \cdot \\ \cdot \\ y_k \end{pmatrix} = \begin{pmatrix} y_1 \\ y_2 \\ \cdot \\ \cdot \\ \cdot \\ y_k \end{pmatrix} + \begin{pmatrix} M_1 \\ M_2 \\ \cdot \\ \cdot \\ \cdot \\ M_k \end{pmatrix} \times x$$

and to self-schedule individual vector operations on each of the blocks in parallel:

$$y_i = y_i + M_i x \text{ for } i = 1, 2, ..., k.$$

That is, the subproblem indexed by i is picked up by a processor as it becomes available, and the entire matrix-vector product is reported done when all of these subproblems have been completed.

If the parallel machine has vector capabilities on each of the processors, this partitioning introduces shorter vectors and defeats the potential of the vector capabilities for small- to medium-size matrices. A better way to partition in this case is

$$y = y + (\hat{M}_1, \hat{M}_2, ..., \hat{M}_k) \begin{pmatrix} x_1 \\ x_2 \\ \cdot \\ \cdot \\ \cdot \\ x_k \end{pmatrix} .$$

Again, subproblems are computed by individual processors. However, in this scheme, we must either synchronize the contribution of adding in each term $\hat{M}_i x_i$ or write each of these into temporary locations and hold them until all are complete before adding them to get the final result.

This scheme does prove to be effective for increasing the performance of the factorization subroutines on the smaller (order less than 100) matrices. During the final stages of the LU factorization, vector lengths become short regardless of matrix size. For the smaller matrices, subproblems with vector lengths that are below a certain performance level represent a larger percentage of the calculation. This problem is magnified when the rowwise partitioning is used.

5.6.2 Matrix-Matrix Product

Matrix-matrix operations offer the proper level of modularity for performance and transportability across a wide range of computer architectures including parallel machines with memory hierarchy. This enhanced performance is primarily due to a greater opportunity for reuse of data. There are numerous ways to accomplish this reuse of data to reduce memory traffic and to increase the ratio of floating-point operations to data movement. Here we present one way.

Consider the operation

$$Y \leftarrow Y + MX,$$

where Y is an $m \times n$ matrix, M is an $m \times p$ matrix, and X is a $p \times n$ matrix. Let us assume that $p + 1$ vector registers are available (typically $p + 1 = 8$). The idea will be to hold p columns of M in p vector registers and successively compute the columns of Y:

$$Y(1:m,j) = Y(1:m,j) + sum_{k=1}^p M(1:m,k)X(k,j) \text{ for } j = 1, 2, ..., n. \tag{5.13}$$

One register is used to hold successive columns of Y. As the computation proceeds across the n columns, there are asymptotically $2p$ flops per 2 vector data accesses (reading and writing the jth column of Y).

If a cache is involved and we can rely on p columns of M remaining in cache once they are referenced, then it might be advantageous to set p to the length of a vector register (typically 32 or 64 words). In this case one would achieve a ratio of $2p$ flops per 3 vector data accesses (reading and writing the jth column of Y and reading the jth column of X), assuming that the rate of access to data in cache is comparable to that in registers.

Within this scheme, blocking of M in order to maintain a favorable cache *hit ratio* is important. We may exploit parallelism through blocking Y and M by rows and assigning each block to a separate processor; this requires only a *fork-join* synchronization. Alternatively, one may block Y and X by columns and assign each of these blocks to a separate processor. As in the matrix-vector operation, either fine-grain synchronization or temporary storage will be required when the blocking for parallelism is by columns.

5.6.3   Typical Performance for Parallel Processing

Blocked algorithms can exploit the parallel-processing capabilities of many high-performance machines, as described in the previous section. The basic idea is to leave the blocked algorithms unchanged and to introduce parallel processing at the Level 3 BLAS layer. The Level 3 BLAS offer greater scope than the Level 1 or 2 BLAS for exploiting parallel processors on shared-memory machines, since more operations are performed in each call.

Tables 5.7 and 5.8 give results for the CRAY Y-MP/8 and CRAY-2/4 for various numbers of processors when parallel processing is being exploited at the Level 3 BLAS layer only. Here the blocksize is 64 for LU decomposition and 16 for the QR factorization. The maximum speed of a single processor of a CRAY Y-MP is 333 Mflops. Thus, we see that for large-enough matrix dimensions, the single-processor code runs at 90% efficiency. When all 8 processors are used, the code attains 73% to 80% efficiency.

Table 5.7: **Timing Results for a CRAY Y-MP/8 (Mflops)**

		Matrix size n					
Algorithm		32	64	128	256	512	1024
LU	(1 proc)	40	108	195	260	290	304
	(2 proc)	32	91	229	408	532	588
	(4 proc)	32	90	260	588	914	1097
	(8 proc)	32	90	205	375	1039	1974
QR	(1 proc)	54	139	225	275	294	301
	(2 proc)	50	134	256	391	505	562
	(4 proc)	50	136	292	612	891	1060
	(8 proc)	50	133	328	807	1476	1937

Table 5.8: **Timing Results for a CRAY 2-S/4 (Mflops)**

		Matrix size n					
Algorithm		32	64	128	256	512	1024
LU	(1 proc)	28	81	152	238	320	381
	(2 proc)	17	54	145	327	540	715
	(4 proc)	17	55	160	394	865	1290
QR	(1 proc)	43	115	195	211	258	265
	(2 proc)	35	98	195	385	576	689
	(4 proc)	35	97	220	528	941	1122

As can be seen, using parallel processing for small-order matrices results in a "slow down" in performance over the single-processor run. For sufficiently large matrices, however, the gain can be substantial—approximately 2 Gflops.

5.6.4   Benefits

Abundant evidence has already been given to verify the viability of these schemes for a variety of parallel and vector architectures. In addition to the computational evidence, several factors support the use of a modular approach. We can easily construct the standard algorithms in linear algebra from these types of module. The operations are simple and yet encompass enough computation that they can be vectorized and also parallelized at a reasonable level of granularity [43]. Finally, the modules can be constructed in such a way that they hide all of the machine intrinsics required to invoke parallel computation, thereby shielding the user from being concerned with any machine-specific changes to the library.

5.7 LAPACK

A collaborative effort is under way to develop a transportable linear algebra library in Fortran 77 [6]. The library is intended to provide a uniform set of subroutines to solve the most common linear algebra problems and to run efficiently on a wide range of high-performance computers.

Specifically, the LAPACK library (shorthand for Linear Algebra Package) will provide routines for solving systems of simultaneous linear equations, least-squares solutions of overdetermined systems of equations, and eigenvalue problems. The associated matrix factorizations (LU, Cholesky, QR, SVD, Schur, generalized Schur) will also be provided, as will related computations such as reordering of the factorizations and condition numbers (or estimates thereof). Dense and banded matrices will be provided for, but not general sparse matrices. In all areas, similar functionality will be provided for real and complex matrices.

The new library will be based on the successful EISPACK [148, 76] and LINPACK [30] libraries, integrating the two sets of algorithms into a unified, systematic library. A great deal of effort is also beeing expended to incorporate design methodologies and algorithms that make the LAPACK codes more appropriate for today's high-performance architectures. The LINPACK and EISPACK codes were written in a fashion that, for the most part, ignored the cost of data movement. As seen in Chapters 3 and 4, most of today's high-performance machines, however, incorporate a memory hierarchy [28, 99, 152] to even out the difference in speed of memory accesses and vectorized floating-point operations. As a result, codes must be careful about reusing data in order not to run at memory speed instead of floating-point speed. LAPACK codes are being carefully restructured to reuse as much data as possible in order to reduce the cost of data movement. A further improvement is the incorporation of new and improved algorithms for the solution of eigenvalue problems [26, 44].

LAPACK is intended to be efficient and transportable across a wide range of computing environments, with special emphasis on modern high-performance computers. It is expected to improve efficiency in two ways. First, the developers are restructuring most of the algorithms in LINPACK and EISPACK in terms of calls to a small number of extended BLAS each of which implements a block matrix operation such as matrix multiplication, rank-k matrix updates, and the solution of triangular systems. These block operations can be optimized for each architecture, but the numerical algorithms that call them will be portable. Second, the developers will implement a class of recent divide-and-conquer algorithms for various eigenvalue problems [44, 102, 8].

The target machines are high-performance architectures with one or more processors, usually with a vector-processing facility. The class includes all of the most powerful SIMD and MIMD machines currently available and in use for general-purpose scientific computing: Alliant FX, Convex, CRAY-2, CRAY X-MP, CRAY Y-MP, Fujitsu VP, Hitachi S-820, IBM 3090/VF, and NEC SX. It is assumed that the number of processors is modest (no more than 100, say). In cases where an algorithm can be restructured into block form in several different ways, all with different performance characteristics, the LAPACK project will choose the structure that provides the best "average" performance over the range of target machines. It is also hoped that the library will perform well on a wider class of shared-memory machines, and the developers are actively exploring the applicability of their approach to distributed-memory machines.

LAPACK will also contain a variety of linear algebra algorithms that are much more accurate than their predecessors. In general, they replace absolute error bounds (either on the backward or forward error) with relative error bounds, and hence better respect the sparsity and scaling structure of the original problems.

The library will be in the public domain, freely distributed over *netlib* [35], just like LINPACK, EISPACK, and many other libraries. The library is expected to be available in 1991.

Chapter 6

Direct Solution of Sparse Linear Systems

In this chapter, we discuss the direct solution of linear equations, where the coefficient matrix is large and sparse. Specifically, we examine algorithms based on Gaussian elimination for solving the equation

$$Ax = b, \tag{6.1}$$

where the coefficient matrix A is large and sparse.

Some of the algorithms may appear quite complicated, but it is important to remember that we are concerned only with direct methods based on Gaussian elimination. That is, all our algorithms compute an LU factorization of a permutation of the coefficient matrix A, so that $PAQ = LU$, where P and Q are permutation matrices, and L and U are lower and upper triangular matrices, respectively. These factors are then used to solve the system (6.1) through the forward substitution $Ly = P^T b$ followed by the back substitution $U(Q^T x) = y$. When A is symmetric, this fact is reflected in the factors, and the decomposition becomes $PAP^T = LL^T$ (Cholesky factorization) or $PAP^T = LDL^T$ (needed for an indefinite matrix). This last decomposition is sometimes called root-free Cholesky. It is common to store the inverse of D rather than D itself in order to avoid divisions when using the factors to solve linear systems. Note that we have used the same symbol L in all three factorizations although they each represent a different lower triangular matrix.

The study of algorithms for effecting such solution schemes when the matrix A is large and sparse is important not only for the problem in its own right, but also because the type of computation required makes this an ideal paradigm for large-scale scientific computing in general. In other words, a study of direct methods for sparse systems encapsulates many issues that appear widely in computational science and that are not so tractable in the context of really large scientific codes.

The principal issues can be summarized as follows:

1. Floating-point calculations themselves form only a small proportion of the total code.

2. The data-handling problem is significant.

3. Storage is often a limiting factor, and auxiliary storage is frequently used.

4. Although the innermost loops are often well defined, usually a significant amount of time is spent in computations in other parts of the code.

5. The innermost loops can sometimes be very complicated.

Issues 1–3 are related to the manipulation of sparse data structures. The efficient implementation of techniques for handling these are of crucial importance in the solution of sparse matrices, and we discuss these in Section 6.1. Similar issues arise when handling large amounts of data in other large-scale scientific computing problems. Issues 2 and 4 serve to indicate the sharp contrast between sparse and non-sparse linear algebra. In code for large full systems, well over 90 percent of the time (on a serial machine) is typically spent in the innermost loop, whereas a substantially lower fraction is spent in the innermost loops of sparse codes. The lack of dominance of a single loop is also characteristic of a wide range of large-scale applications.

Specifically, the data handling nearly always involves indirect addressing (see Section 3.5). This problem clearly has implications for vector and parallel architectures, particularly if the hardware gather/scatter is to be used efficiently.

Another way in which the solution of sparse systems acts as a paradigm for a wider range of scientific computation is that it exhibits a hierarchy of parallelism that can be used to good account in many sparse matrix algorithms. This hierarchy comprises three levels:

- *System level.* This involves the underlying problem which, for example, may be a partial differential equation (PDE) or a large structures problem. In these cases, it is natural (perhaps even before discretization) to subdivide the problem into smaller subproblems, solve these independently, and combine the independent solutions through a small (usually dense) interconnecting problem. In the PDE case, this is done through domain decomposition; in the structures case, it is called substructuring. An analogue in discrete linear algebra is partitioning and tearing.

- *Matrix level.* At this level, parallelism is present because of sparsity in the matrix. A simple example lies in the solution of tridiagonal systems. In a structural sense no (direct) connection exists between variable 1 in equation 1 and variable n in equation n (the system is assumed to have order n), and so Gaussian elimination can be applied to both of these "pivots" simultaneously. The elimination can proceed, pivoting on entry 2 and $n - 1$, then 3 and $n - 2$ simultaneously so that the resulting factorization (known in LINPACK as the BABE algorithm) has a parallelism of two. This is not very exciting, although the amount of arithmetic is unchanged from the normal sequential factorization. However, sparsity allows us to pivot simultaneously on every other entry; and when this is continued in a nested fashion, the cyclic reduction (or nested dissection in this case also) algorithm results. Now we have only $\log n$ parallel steps, although the amount of arithmetic is about double that of the serial algorithm. This sparsity parallelism can be automatically exploited for any sparse matrix, as we discuss in Section 6.5.

- *Submatrix level.* This level is exploited in an identical way to Chapter 5, since we are here concerned with eliminations within full submatrices of the overall sparse system. Thus, the techniques of the full linear algebra case (e.g., Level 3 BLAS) can be used. The only problem is how to organize the sparse computation to yield operations on full submatrices. This is easy with band, variable band, or frontal solvers (Section 6.4) but can also be extended, through multifrontal methods, to any sparse system (Section 6.5).

6.1   Introduction to Direct Methods for Sparse Linear Systems

The methods that we consider for the solution of sparse linear equations can be grouped into three main categories: general techniques, frontal methods, and multifrontal approaches. We go into some detail on the methods because we wish to draw attention to the features that are important in exploiting vector and parallel architectures. For further background on these techniques, we recommend the book by Duff, Erisman, and Reid [56]. In this section we introduce the algorithms and approaches and examine some basic operations on sparse matrices.

6.1.1   Three Approaches

The study of sparse matrix techniques is in great part empirical, so throughout we illustrate points by reference to runs of actual codes. We discuss the availability of codes in the penultimate section of this chapter.

We first consider (Section 6.2) a completely *general approach* typified by the Harwell Subroutine MA28 [47] or Y12M [173]. The principal features of this general approach are that numerical and sparsity pivoting are performed at the same time (so that dynamic data structures are used in the initial factorization) and that sparse data structures are used throughout—even in the inner loops. As we shall see, these features must be considered drawbacks with respect to vectorization and parallelism. The strength of the general approach is that it will give a very satisfactory performance over a wide range of structures. Some gains and simplification can be obtained if the matrix is symmetric or banded. We discuss these methods and algorithms in Section 6.3.

Frontal schemes can be regarded as an extension of band or variable-band schemes and will perform well on systems whose bandwidth or profile is small. The efficiency of such methods for solving grid-based problems (for example, discretizations of partial differential equations) will depend crucially on the underlying geometry of the problem. One can, however, write frontal codes so that any system can be solved; sparsity preservation is obtained from an initial ordering, and numerical pivoting can be performed within this ordering. A characteristic of frontal methods is that no indirect addressing is required in the innermost loop. We use the code MA32 from the Harwell Subroutine Library [97] to illustrate this approach in Section 6.4.

The final class of techniques, which we study in Section 6.5, is an extension of the frontal methods termed *multifrontal*. The extension permits efficiency for any matrix whose nonzero pattern is symmetric or nearly symmetric and allows any sparsity ordering techniques for symmetric systems to be used. The restriction to nearly symmetric patterns arises because the initial ordering is performed on the sparsity pattern of the Boolean sum of A and A^T. The approach can, however, be used on any system (for example, Harwell Subroutine Library code MA37 [63]). As in the frontal method, full matrices are used in the innermost loop so that indirect addressing is avoided. There is, however, more data movement than in the frontal scheme, and the innermost loop is not so dominant. In addition to the use of direct addressing, multifrontal methods also differ from the first class of methods because the sparsity pivoting is separated from the numerical pivoting.

6.1.2   Description of Sparse Data Structure

We continue this introductory section by describing the most common sparse data structure, which is the one used in most general-purpose codes. The structure for a row of the sparse matrix is illustrated in Figure 6.1. All rows are stored in the same way, and a pointer is used to identify the location of the beginning of the data structure for each row. If the pointer was that for row i, then entry a_{i,j_1} would have value ξ. Clearly, access to the entries in a row is straightforward, although indirect addressing is required to identify the column of an entry.

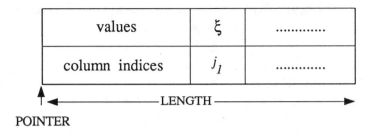

Figure 6.1: **General sparse matrix storage scheme**

We illustrate this scheme in detail using the example in Figure 6.2.

$$
\begin{pmatrix}
1 & 0 & 0 & 4 \\
-1 & 0 & 3 & 0 \\
0 & 2 & 0 & 0 \\
0 & -2 & 0 & -4
\end{pmatrix}
$$

Figure 6.2: **A** 4×4 **sparse matrix**

If we consider the matrix in Figure 6.2, we can hold each row as a packed sparse vector and the matrix as a collection of such vectors. For each member of the collection we normally store a pointer to its start and the number of entries. Since we are thinking in terms of Fortran 77, a pointer is simply an array subscript indicating a position in an array. Thus, for example, the matrix of Figure 6.2 may be stored in Fortran arrays as shown in Table 6.1. Here LEN(i) contains the number of entries in row i, while IPTR(i) contains the location in arrays ICN and VALUE of the first entry in row i. For example, row 2 starts in position 3; referring to position 3 in ICN and VALUE, we find the (2,3) entry has value 3. Since LEN(2) = 2, the fourth position is also in row 2; specifically the (2,1) entry has value -1. A detailed discussion of this storage scheme is given in [65].

Table 6.1: **Matrix of Figure 6.2 Stored as a Collection of Sparse Row Vectors**

Subscripts	1	2	3	4	5	6	7
LEN	2	2	1	2			
IPTR	1	3	5	6			
ICN	4	1	3	1	2	2	4
VALUE	4.	1.	3.	-1.	2.	-2.	-4.

Note that there is a certain redundancy in holding both row lengths (LEN) and row pointers (IPTR) when the rows are held in order, as in Table 6.1. In general, however, operations on the matrix will result in the rows not being in order and thus will require both arrays.

6.1.3 Manipulation of Sparse Data Structure

To give a flavor of the issues involved in the design of sparse matrix algorithms, we now examine a manipulation of sparse data structures that occurs commonly in LU factorization. The particular manipulation we consider is the addition of a multiple of one row (the pivot row) of the matrix to another row (the nonpivot row) where the matrix is stored in the row pointer/column index scheme just described. Assume that an integer array, IQ say, containing all positive entries is available and that there is sufficient space to hold temporarily two copies of the pivot row in sparse format. Assume that the second copy has been made. Then a possible scheme is as follows:

1. Scan the pivot row, and for each entry determine its column from the ICN value. Set the corresponding entry in IQ to the negation of the position of that entry within the compact form of the pivot row. Hold the original IQ entry in the appropriate position in ICN.

   For each nonpivot row, do steps 2 and 3:

2. Scan the nonpivot row. For each column, check the corresponding entry in IQ. If it is negative, set it positive, and update the value of the appropriate entry in the nonpivot row. If it is positive, continue to the next entry in the nonpivot row.

3. Scan the unaltered copy of the pivot row. If the corresponding IQ value is positive, set it negative. If it is negative, then there is fill-in to the nonpivot row. The new entry is added, and we continue to the next entry in the pivot row.

4. Finally, reset IQ using information from both copies of the pivot row.

Figure 6.3 illustrates the situation before each step. In this figure, the upper row for IQ gives the contents of the locations indicated in the lower row. Note that it is not necessary to keep each row in column order when using this scheme. Conceptually simpler sparse vector additions can result when they are kept in order, but the scanning of sparse arrays in parallel and the extra work to keep the columns in order can lead to inefficiencies.

Before (i)

IQ
$$\begin{array}{ccccc} i_1 & i_2 & i_3 & i_4 & i_5 \\ j_2 & j_1 & j_4 & j_3 & j_5 \end{array}$$

Pivot row
$$\begin{array}{c} A \\ ICN \end{array} \quad \begin{array}{ccc} \alpha_1 & \alpha_2 & \alpha_3 \\ j_1 & j_2 & j_3 \end{array} \quad + \quad \text{second copy}$$

Non-pivot row
$$\begin{array}{c} A \\ ICN \end{array} \quad \begin{array}{cccc} \beta_1 & \beta_2 & \beta_3 & \beta_4 \\ j_2 & j_3 & j_4 & j_5 \end{array}$$

Before (ii)

IQ
$$\begin{array}{ccccc} -2 & -1 & i_3 & -3 & i_5 \\ j_2 & j_1 & j_4 & j_3 & j_5 \end{array}$$

Pivot row
$$\begin{array}{c} A \\ ICN \end{array} \quad \begin{array}{ccc} \alpha_1 & \alpha_2 & \alpha_3 \\ i_2 & i_1 & i_4 \end{array} \quad \text{and} \quad \begin{array}{ccc} \alpha_1 & \alpha_2 & \alpha_3 \\ j_1 & j_2 & j_3 \end{array} \quad \text{stored separately}$$

No change to non-pivot row

Before (iii)

IQ
$$\begin{array}{ccccc} 2 & -1 & i_3 & 3 & i_5 \\ j_2 & j_1 & j_4 & j_3 & j_5 \end{array}$$

Pivot row unchanged

Non-pivot row
$$\begin{array}{c} A \\ ICN \end{array} \quad \begin{array}{cccc} \beta_1 + \zeta\alpha_2 & \beta_2 + \zeta\alpha_3 & \beta_3 & \beta_4 \\ j_2 & j_3 & j_4 & j_5 \end{array}$$

Before (iv)

IQ
$$\begin{array}{ccccc} -2 & -1 & i_3 & -3 & i_5 \\ j_2 & j_1 & j_4 & j_3 & j_5 \end{array}$$

Pivot row unchanged

Non-pivot row
$$\begin{array}{c} A \\ ICN \end{array} \quad \begin{array}{ccccc} \beta_1 + \zeta\alpha_2 & \beta_2 + \zeta\alpha_3 & \beta_3 & \beta_4 & \zeta\alpha_1 \\ j_2 & j_3 & j_4 & j_5 & j_1 \end{array}$$

Figure 6.3: **Sketch of steps involved when adding one sparse vector to another**

The important point about the rather complicated algorithm just described is that there are no scans of length n vectors. Thus, if the computation is performed at each major step of Gaussian elimination on a matrix of order n, there are (from this source) no $O(n^2)$ contributions to the overall work. Since our target in sparse calculations is to develop algorithms that run as $O(\tau) + O(n)$, for a matrix of order n with τ nonzeros, $O(n^2)$ calculations are a disaster and will dominate the computation if n is large enough.

In addition to avoiding such scans, this scheme permits the indices to be held in any order, and no REAL vector of length n is needed. Indeed, the array IQ in Figure 6.3 can be used for any other purpose; our only assumption was that its entries were all positive. Another benefit is that we never check numerical values for zero (which might be the case if we expanded into a full-length REAL vector), so no confusion arises between explicitly held zeros and zeros not within the sparsity pattern. This point can be important when solving several systems with the same structure but different numerical values.

6.2   General Sparse Matrix Methods

A major concern when the matrix A is sparse is that the factors L and U will generally be denser than the original A.

Fill-in is caused in Gaussian elimination if, in the basic operation

$$a_{ij} \leftarrow a_{ij} - a_{ik} a_{kk}^{-1} a_{kj}, \tag{6.2}$$

the entry in location (i,j) of the original A was zero. The ordering of A can be important in preserving sparsity in the factors. Figure 6.4 gives an example of a case where ordering the rows and columns to preserve sparsity in Gaussian elimination is extremely effective. If pivots are chosen from the diagonal in the natural order, the reordered matrix preserves all zeros in the factorization, but the original order preserves none.

Table 6.2 illustrates the gains one can obtain in the sparse case by ignoring all or most of the zero entries both in the original matrix and in the ensuing calculation. The second row of the table gives figures for sparse elimination without any reordering, while the third row indicates that further substantial gains can be made when sparsity-exploiting orderings are used. In larger problems, we might expect even more significant gains, since storage and work for Gaussian elimination on a full system of order n behave as $O(n^2)$ and $O(n^3)$, respectively.

Another difference between full and sparse systems arises when we consider the common case where a subsequent solution is required with a matrix of the same sparsity structure as a previously factored system. Whereas this has no relevance in the full case, it does have a considerable influence

```
x x x x x x x                    x              x
x x                                 x           x
x   x                                  x        x
x     x                                   x     x
x       x                                    x  x
x         x                                     x  x
x           x                                      x x
x             x                  x x x x x x x x
```

Original matrix   Reordered matrix

Figure 6.4: **Original and reordered matrix**

Table 6.2: **Benefits of Sparsity on Matrix of Order 199 with 873 Nonzeros**

Procedure	Total storage (Kwords)	Thousands of flops	Time in sec. on an IBM 3033
Treating system as full	39.6	5254	3.382
Storing and operating only on nonzeros	9.3	254	.211
Using sparsity pivoting	3.3	8	.019

in the sparse case since information from the first factorization can be used to simplify the second. Indeed, in some instances, most or all of the ordering and data organization can be done before any numerical factorization is performed.

However, a much more significant difference is that the computed inverse of an irreducible sparse matrix is always full [55], whereas the factors can often be very sparse. Examples are a tridiagonal matrix and the arrowhead matrix shown in Figure 6.4. In the full case, for stability and work considerations, one may argue for using triangular factors rather than explicit inverses; in the sparse case, however, the savings from using sparse factors is what makes a direct factorization approach feasible. In short, it is important that one does not use explicit inverses when dealing with large sparse matrices.

We illustrated the effect of ordering on sparsity preservation in Figure 6.4. A simple but effective strategy for maintaining sparsity is due to Markowitz [121]. At each stage of Gaussian elimination, he selects as a pivot the nonzero entry of the remaining reduced submatrix with the lowest product of number of other entries in its row and number of other entries in its column.

More precisely, after the kth major step of Gaussian elimination, let $r_i^{(k)}$ denote the number of entries in row i of the reduced $(n-k) \times (n-k)$ submatrix, Similarly let $c_j^{(k)}$ be the number of entries in column j. The Markowitz criterion chooses the entry $a_{ij}^{(k)}$ from the reduced submatrix to minimize the expression

$$(r_i^{(k)} - 1)(c_j^{(k)} - 1), \tag{6.3}$$

where $a_{ij}^{(k)}$ satisfies some numerical criteria also.

This strategy can be interpreted in several ways, for example, as choosing the pivot to modify the least number of coefficients in the remaining submatrix. It may also be regarded as an attempt to reduce the number of multiplications, since using $a_{ij}^{(k)}$ as pivot requires $r_i^{(k)}(c_j^{(k)} - 1)$ multiplications. Finally, we may think of (6.3) as a measure of the fill-in at this stage, since it would be so if all $(r_i^{(k)} - 1)(c_j^{(k)} - 1)$ modified entries were previously zero.

In general, for the Markowitz ordering strategy in the unsymmetric case, we need to establish a suitable control for numerical stability. In particular, we restrict the Markowitz selection to those pivot candidates satisfying the inequality

$$a_{kk}^{(k)} \geq u a_{ik}^{(k)}, \quad i \geq k, \tag{6.4}$$

where u is a preset threshold parameter in the range $0 < u \leq 1$.

If we look back at equation (6.2), we see that the effect of u is to restrict the maximum possible growth to $(1 + 1/u)$. Since it is possible to relate the entries of the backward error matrix E (that is, the LU factors are exact for the matrix $A + E$ to such growth by the formula

$$|e_{ij}| \leq 3.01 \times \epsilon \times n \times \max a$$

where ϵ is the machine precision and $\max a$ the largest entry encountered in the Gaussian elimination process), then changing u can affect the stability of our factorization. We illustrate this effect in Table 6.3 and remark that, in practice, a value of u of 0.1 has been found to provide a good compromise between maintaining stability and having the freedom to reorder to preserve sparsity.

Table 6.3: **Effect of Variation in Threshold Parameter** u

(matrix of order 541 with 4285 nonzeros)

u	Nonzeros in Factors	Error
1.0	16767	3×10^{-9}
0.25	14249	6×10^{-10}
0.10	13660	4×10^{-9}
0.01	15045	1×10^{-5}
10^{-4}	16198	1×10^{2}
10^{-10}	16553	3×10^{23}

The results in Table 6.3 are without any iterative refinement, but even with such a device no meaningful answer is obtained in the case with u equal to 10^{-10}. If in the current vogue for inexact factorizations, we consent to sacrifice accuracy of factorization for increase in sparsity, then this is done not through the threshold parameter u but rather through a drop tolerance parameter *tol*. Entries encountered during the factorization of value less than *tol* or less than *tol* times the largest in the row or column are dropped from the structure, and an inexact or partial factorization of the matrix is obtained.

Most recent vector processors have a facility for hardware indirect addressing, and one might be tempted to believe that all problems associated with indirect addressing have been overcome. For example, on the CRAY X-MP, the loop shown in Figure 6.5 (a sparse SAXPY) ran asymptotically at only 5.5 Mflops when hardware indirect addressing was inhibited but ran asymptotically at over 80 Mflops when it was not. (For further information on the behavior of a sparse SAXPY on a range of machines, see Section 4.5.)

```
        DO 100 I=1,M
            A(ICN(I)) = A(ICN(I)) + AMULT * W(I)
    100 CONTINUE
```

Figure 6.5: **Sparse SAXPY loop**

On the surface, the manufacturers' claims to have conquered the indirect addressing problem would seem vindicated, and we might believe that our sparse general codes would perform at about half the rate of highly tuned full matrix code. This reasoning has two flaws. The first lies in the $n_{1/2}$ value [94] for the sparse loops (i.e., the length of the loop required to attain half the maximum performance). This measure is directly related to the startup time. For the loop shown in Figure

Table 6.4: **Statistics from MA28 on an IBM 3084
for Various Matrices from Different Disciplines**

Order of Matrix	No. of Entries	Area of Study	Ave. Length of Pivot Row	% Time in Inner Loops
1454	3377	Power systems networks	2	39
1176	9874	Electronic circuit analysis	8	58
1107	5664	Computer simulation	19	42
1005	4813	Ship design	24	50
838	5424	Aerospace	27	49
655	2854	Chemical engineering	3	37
541	4285	Atmospheric pollution	10	52
420	7252	Structural analysis	29	49
363	3157	Linear programming	5	35
199	701	Stress analysis	3	43
147	2449	Atomic spectra	17	55

6.5, the $n_{1/2}$ value on the CRAY X-MP is about 50, which—relative to the typical order of sparse matrices being solved by direct methods (greater than 10,000)—is insignificant. However, the loop length for sparse calculations depends not on the order of the system but rather on the number of entries in the pivot row. We have done an extensive empirical examination of this number using the MA28 code on a wide range of applications. We show a representative sample of our results in Table 6.4. Except in the examples from the analysis of structures (the last three in the table), this length is very low and, even in these small structures examples, is much less than the $n_{1/2}$ value mentioned above. Thus the typical performance rate for the sparse inner loop is far from the asymptotic performance.

It should be added that there are matrices where the number of entries in each row is very high, or becomes so after fill-in. This would be the case in large structures problems and for most discretizations of elliptic partial differential equations. For example, the factors of the five-diagonal matrix from a five-point star discretization of the Laplace operator on a q by q grid could have about $6 \log_2 q$ entries in each row of the factors. Unfortunately, such matrices are very easy to generate and are thus sometimes overused in numerical experiments.

The second problem with the use of hardware indirect addressing in general sparse codes is that the amount of data manipulation in such a code means that a much lower proportion of the time is spent in the third level of loops than in code for dense matrices. Again we have performed an empirical study on MA28; we show these results in the last column of Table 6.4. The percentage given in that table is for the total time of three loops in MA28, all at the same depth of nesting. We see that typically around 50 percent of the overall time on an IBM 3084 is spent in the innermost loops. Thus, even if these loops were made to run infinitely fast, a speedup of only about a factor of two would be obtained: a good illustration of Amdahl's law.

Of course, if we can avoid the necessity for numerical pivoting, then the code and the inner-loop structure can be made much simpler. For example, code for positive definite systems [80] has been modified by Boeing Computer Services so that it uses hardware indirect addressing on the CRAY X-MP. On typical problems, they have decreased the execution time by a factor of nearly four. One should mention, however, that more recent work effectively blocks the operations so that the indirect addressing is removed from the inner loops. If this approach (similar in many ways to the one we discuss in Section 6.5) is used, much higher computational rates can be achieved.

We conclude, therefore, that for general matrices vector indirect addressing is of limited assistance for present-generation general sparse codes. Even for general systems, however, advantage can be taken of vectorization by using a hybrid approach, where a full matrix routine is used when fill-in has caused the reduced matrix to become sufficiently dense. That is, at some stage, it is not worth paying attention to sparsity. At this point, the reduced matrix can be expanded as a full matrix, holding any remaining zero entries explicitly, and a full matrix code can then be used to effect the subsequent decomposition. The resultant hybrid code should combine the advantages of the reduction in operations resulting from sparsity in the early stages with the high computational rates of a full linear algebra code in the latter. The point at which such a switch is best made will, of course, depend both on the vector computer characteristics and on the relative efficiency of the sparse and full codes.

Experiments with such an approach on the CRAY-1 indicate that the switchover density for overall time minimization can often be low (typically 20 percent dense) and that gains of a factor of over four can be obtained even with unsophisticated full matrix code [50]. In Table 6.5, we report on the performance of a more recent version of the hybrid code on the CRAY-2, where the sparse decomposition is performed by MA28 (with threshold parameter 0.1) and the full decomposition by the assembler-coded full matrix solver from the Cray Research Scientific Library SCILIB (with the same name, calling sequence, and functionality of the full LU factorization routine SGEFA from LINPACK).

Table 6.5: **CRAY-2 Performance of MA28 with Switch to Full Code**

(Matrix from five-point discretization of the Laplacian on a 50×50 grid)

Density for Switch to Full Code	Order of "Full" Matrix	Millions of Operations	Time in Seconds
No switch	0	7	21.8
1.00	74	7	21.4
0.80	190	8	15.0
0.60	235	11	12.5
0.40	305	21	9.0
0.20	422	50	5.5
0.10	531	100	3.7
0.05	677	207	2.7
0.03	804	346	2.6
0.01	1182	1100	3.9
0.005	1420	1908	6.1

Several comments can be made on these results. First, the speedup of over 8 is substantial and far better than on the CRAY-1. Second, any suggestion of abandoning the use of sparse codes is refuted since it is clear that the increase in number of operations by switching at lower and lower densities of reduced matrix eventually more than compensates for the much higher computational rate of the full code. Indeed, for this matrix of order only 2500, it is best to perform more than half the eliminations in sparse mode; the solution with the SCILIB version of SGEFA on the whole system required over 30 seconds. Additionally, the example we have used is a particularly bad one for a sparse direct code since fill-in is high for such regular problems. On a network problem, for example, the reduced matrix remains quite sparse until very late in the factorization.

Naturally, if we store the zeros of a reduced matrix, we might expect an increase in the storage requirements for the decomposition. Although the luxury of 256 Mwords on the CRAY-2 gives us ample scope for allowing such an increase, it is interesting to observe that the increase in storage for floating-point numbers is to some extent compensated for by the reduction in storage for the accompanying integer index information. Thus, in our example, the overall storage requirements increase from 300,120 words for the entirely sparse code to 327,988 words at a switchover density of 0.6, and to 691,153 words at a switchover density of 0.1. Admittedly, at the optimal switchover in terms of speed, about 1.25 million words are required; and, for the run in the last row of Table 6.5, over 4 million words of storage were needed.

Another observation (not recorded in Table 6.5) is that, although the number of operations increases dramatically, the accuracy of the decomposition (reflected in the residuals for the solution of the equations) is not adversely affected. Indeed, the greater stability of using partial pivoting (effectively, a threshold of 1.0), as opposed to threshold pivoting with a threshold of 0.1, yields a residual norm of 1.4×10^{-10} at the optimal switchover as against a residual norm of 1.1×10^{-8} for the original sparse code. A higher value of the threshold would, of course, give greater stability to the sparse code at the cost of greater fill-in and an earlier switch to the full code option.

In this section, we have concentrated on the approach typified by the codes MA28 [47] and Y12M [173]. While these are possibly the most common codes used to solve general systems arising in a wide range of application areas, there are other algorithmic approaches and codes for solving general unsymmetric systems. One technique is to preorder the rows, then to choose pivots from each row in turn, first updating the incoming row according to the previous pivot steps and then choosing the pivot by using a threshold criterion on the appropriate part of the updated row. If the columns are also preordered for sparsity, then an attempt is first made to see whether the diagonal entry in the reordered form is suitable. This approach is used by the codes NSPFAC and NSPIV of Sherman [145]. It is similar to factorizations subsequent to the initial one using the main approach of this section, and so is very simple to code. It can, however, suffer badly from fill-in if a good initial ordering is not given or if numerical pivoting forbids keeping close to that ordering.

Another approach is to generate a data structure that, within a chosen column ordering, accommodates all possible pivot choices [83]. It is remarkable that this is sometimes not overly expensive in storage and has the benefit of good stability but within a subsequent static data structure. There are, of course, cases when the storage penalty is high.

Methods using a sparse QR factorization can be used for general unsymmetric systems. These are based on work of George and Heath [79] and first obtain the structure of R through a symbolic factorization of the structure of the normal equations matrix $A^T A$. It is common not to keep Q but to solve the system using the semi-normal equations

$$R^T R x = A^T b.$$

A QR factorization can, of course, be used for least-squares problems and implementations have been developed using ideas similar to those discussed later in Section 6.5 for Gaussian elimination. Further discussion of QR and least squares is outside the scope of this chapter, which is to discuss techniques based on Gaussian elimination.

All the methods discussed in the preceding paragraphs permit numerical pivoting so that general unsymmetric systems can be stably handled. Here, we do not discuss techniques that have no numerical pivoting; but towards the end of Section 6.3 we shall remark on their benefits and pitfalls, when we have discussed the symmetric case.

We have not discussed the exploitation of parallelism by general sparse direct methods. One possibility is the use of partitioning methods and, in particular, the use of the bordered block triangular form. This approach is discussed by Arioli and Duff [9], who indicate that reasonable speedups on machines with low levels of parallelism (4–8) are obtained fairly easily even on very difficult systems.

6.3   Methods for Symmetric Matrices and Band Systems

If the pattern of A is symmetric and we can be sure that diagonal pivots produce a stable factorization (the most important example is when A is symmetric and positive definite), then two benefits occur. We do not have to carry numerical values to check for stability, and the search for the pivot is simplified to finding i such that

$$r_i^{(k)} = \min_t r_t^{(k)}$$

and using $a_{ii}^{(k)}$ as pivot. This scheme was introduced by Tinney and Walker [155] as their Scheme 2 and is normally termed the *minimum degree algorithm* because of its graph theoretic interpretation: in the graph associated with a symmetric sparse matrix, this strategy corresponds to choosing for the next elimination the node that has the fewest edges connected to it. Surprisingly (as we shall see later in this section), the algorithm can be implemented in the symmetric case without explicitly updating the sparsity pattern at each stage, a situation that greatly improves its performance.

Elementary graph theory has been used as a powerful tool in the analysis and implementation of diagonal pivoting on symmetric matrices (see, for example, [80]). Although graph theory is sometimes overused in sparse Gaussian elimination, in certain instances it is a useful and powerful tool. We shall now explore one such area. For this illustration, we consider finite *undirected graphs*, G(V,E), which comprise a finite set of distinct *vertices* V and an *edge set* E consisting of unordered pairs of vertices. An edge $e \in$ E will be denoted by (u, v) for some $u, v \in$ V. The graph is *labeled* if the vertex set is in 1-1 correspondence with the integers $1, 2, ..., |V|$, where $|V|$ is the number of vertices in V. In this application of graph theory, the set E, by convention, does not include self-loops (edges of the form (u, u), $u \in$ V) or multiple edges. Thus, since the graph is undirected, edge (u, v) is equal to edge (v, u), and only one is held. A subgraph H(U,F) of G(V,E) has vertex set U $\subseteq$ V, and edge set F $\subseteq$ E. H(U,F) is fully connected if $(u, v) \in$ F for all $u, v \in$ U. With any symmetric matrix, of order n say, we can associate a labeled graph with n vertices such that there is an edge between vertex i and vertex j (edge (i, j)) if and only if entry a_{ij} (and, by symmetry, a_{ji}) of the matrix is nonzero. We give an example of a matrix and its associated graph in Figure 6.6.

The main benefits of using such a correspondence can be summarized as follows:

1. The structure of the graph is invariant under symmetric permutations of the matrix (they correspond merely to a relabeling of the vertices).

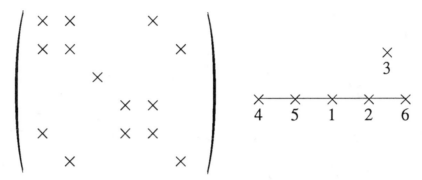

Figure 6.6: **Symmetric matrix and associated graph**

2. For mesh problems, there is usually an equivalence between the mesh and the graph of the resulting matrix. We can thus work more directly with the underlying structure.

3. We can represent *cliques* (fully connected subgraphs) in a graph by listing the vertices in a clique without storing all the interconnection edges.

6.3.1   The Clique Concept in Gaussian Elimination

To illustrate the importance of the clique concept in Gaussian elimination, we show in Figure 6.7 a matrix and its associated graph (also the underlying mesh, if, for example, the matrix represents the five-point discretization of the Laplacian operator). If the circled diagonal entry in the matrix were chosen as pivot (admittedly not a very sensible choice on sparsity grounds), then the resulting reduced matrix would have the dashed (pivot) row and column removed and have additional nonzeros (fill-ins) in the checked positions and their symmetric counterparts. The corresponding changes to the graph cause the removal of the circled vertex and its adjacent edges and the addition of all the dashed edges shown.

Thus, after the elimination of vertex 7, the vertices 3,6,8,11 form a clique, and it is an easily proven property of Gaussian elimination that all vertices of the graph of the reduced matrix that were connected to the vertex associated with the pivot will become pairwise connected after that step of the elimination process. Although this clique formation (and, later in the process, clique amalgamation) has been observed for some time, only recently have techniques exploiting clique formation been used in ordering algorithms. Our example clique has 4 vertices and 6 interconnecting edges; but, as the elimination progresses, this difference will be more marked, since a clique on q vertices has $q(q-1)/2$ edges corresponding to the off-diagonal nonzeros in the associated full

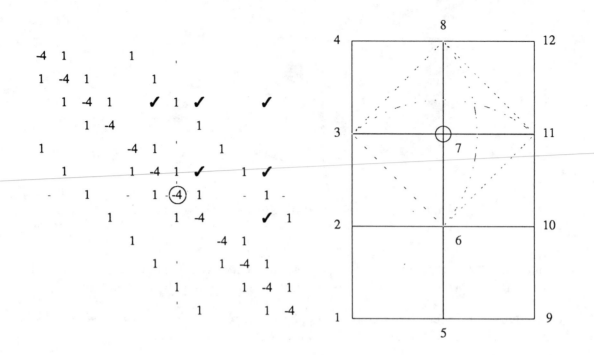

Figure 6.7: **Use of cliques in Gaussian elimination**

submatrix. Since any subsequent clique amalgamation can be performed in a time proportional to the number of edges in the cliques concerned, both work and storage are linear rather than quadratic in the number of nodes in the cliques.

We illustrate clique amalgamation in Figure 6.8 where the circled element is being used as a pivot and the vertices in each rectangle are all pairwise connected. We do not show the edges internal to each rectangle, because we wish to reflect the storage and data manipulation scheme that will be used.

Two cliques (sometimes called elements or generalized elements by analogy with elements in the finite element method) are held only as lists of constituent vertices (1,2,3,4,7,8,10,11,12,13) and (4,5,6,8,9,13,14,15). After the elimination, the variables in both of these cliques will be pairwise connected to form a new clique given by the list (1,2,3,4,5,6,7,9,10,11,12,13,14,15). Not only is a list merge the only operation required, but the storage after elimination (for the single clique) will be less than before the elimination (for the two cliques).

To summarize, two important aspects of such an approach make this development one of the

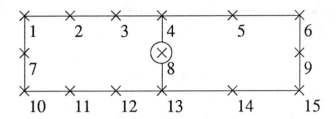

Figure 6.8: **Clique amalgamation**

most exciting in the past decade. First, because we do not have to mimic the Gaussian elimination operations during this ordering phase, the ordering can be significantly faster than the actual elimination. This is in sharp contrast to the situation for unsymmetric systems or indeed to the situation for symmetric systems ten years ago. The second—and in many ways more important— aspect is that of storage, where the ordering can be implemented so that it requires only slightly more storage (a small multiple of n) over that for the matrix itself. Thus, even for very large problems whose decomposition must necessarily be out of core, it may be possible to do an in-core ordering. Additionally, since the storage required for computing the ordering is independent of the fill-in and is known in advance, we can ensure that sufficient storage is available for successful completion of this phase. When computing the ordering we can determine the storage and work required for subsequent factorization. This forecast could be crucial if we wish to know whether the in-core solution by a direct method is feasible or whether an out-of-core solver or iterative method must be used.

To illustrate the remarkable power of current implementations of minimum degree, we show in Table 6.6, from [56], ordering times for various algorithms on three problems arising from successively finer finite-element triangulations of an L-shaped region. Codes MA17A and MA17E are both from the Harwell Subroutine Library and do not use the clique amalgamation concept; however, MA17E uses more efficient internal data structures than MA17A. The Yale Sparse Matrix Package (YSMP), SPARSPAK (University of Waterloo), and MA27A (Harwell Subroutine Library) all use successive refinements of the generalized element model. When we realize that the MA17 codes were considered very powerful in the early 1970s, we see just how dramatic the data in Table 6.6 is.

Table 6.6: **Ordering Times on Graded-L (IBM 370/168, in sec)**

Order	265	1009	3466
Nonzeros	1753	6865	23896
MA17A (1970)	1.56	29.9	≤ 250
MA17E (1973)	0.68	6.86	62.4
YSMP (1978)	0.27	1.11	4.04
MA27A (1981)	0.15	0.58	2.05

Researchers have added numerous twists to the minimum degree scheme. Many of these are implementation refinements; they are discussed in a recent review by George and Liu [81]. Others are developed to compromise a strict minimum degree so that some other property is enhanced—for example, so that the ordering produced retains the good reduction in work afforded by minimum degree while, at the same time, yields an ordering more amenable to parallel implementation than strict minimum degree [57], [116].

Unsymmetric matrices exist for which pivoting down the diagonal in any order is stable and for which, in theory, ordering techniques could be developed to take advantage of the improvements illustrated in Table 6.6. The class of such problems is unknown, however, although both theoretical and practical evidence suggests that matrices arising from the discretization of second-order non-selfadjoint elliptic partial differential equations with constant coefficients fall into this category.

6.3.2   Code Performance and Symmetry

Two main dangers are associated with diagonal pivoting on unsymmetric systems. The most obvious one lies in determining that such pivoting is numerically stable; the second lies in approximating the structure of an unsymmetric system by a symmetric one. Common choices are to use the structure of $A + A^T$ or the symmetric pattern whose upper (or lower) triangle is equivalent to the upper (or lower) triangle of A. These choices could, however, be poor if A is far from symmetric. Additionally, the information from the ordering on work and storage for subsequent factorization may not be at all accurate. Table 6.7 shows the performance of a diagonal pivoting algorithm on two sparse matrices. The matrix of order 541 has a nearly symmetric structure, while that of order 822 is very unsymmetric. These results, from [49], illustrate the dangers of the indiscriminate use of diagonal pivoting on unsymmetric matrices.

Table 6.7 Diagonal Pivoting on Unsymmetric Matrices on an IBM 370/168 in 64-bit Arithmetic

	541	822
Order	541	822
Nonzeros	4285	5607
Nonzeros in factors:		
diagonal	14025	95292
general	13623	6653
Time in sec. on IBM 370/168		
Factorization and sparsity pivoting:		
diagonal	1.98	31.89
general	3.99	14.98
Numerical factorization:		
diagonal	.29	16.18
general	.56	2.72
Solution:		
diagonal	.03	.21
general	.05	.03
l_2 *norm of error:*		
diagonal	8×10^{-6}	8×10^{-9}
general	4×10^{-9}	1×10^{-11}

Similarly, while it is possible to use a general-purpose code on symmetric systems, we would expect the performance to be poor because of the neglect of symmetry. This is borne out by the results from [49] shown in Table 6.8, where the codes MA28 [47] and MA27 [61, 62] are from the Harwell Subroutine Library. The brute-force approach of performing pivot selection and factorization together results in very complicated coding but at the same time yields algorithms of the most general applicability. As one might expect, however, such methods take little account of the underlying structure and so will do rather badly on regularly structured systems.

Table 6.8 **Use of General Code on Symmetric Systems.** The codes used are MA28 (unsymmetric) and MA27 (symmetric).

Order	199	147	292	900
Nonzeros	536	1298	1250	4322
	Time (sec. on IBM 3033)			
Pivot selection and factorization:				
Symmetric code	.09	.12	.17	.94
Unsymmetric code	.20	.60	.58	7.43
Numerical factorization:				
Symmetric code	.04	.08	.09	.64
Unsymmetric code	.05	.17	.14	.96
Storage (words × 1000 on IBM 3033):				
Symmetric code	3.7	6.9	7.6	46.7
Unsymmetric code	7.5	15.0	17.2	108.0

More recently, there has been some work aimed at exploiting vectorization by blocking the calculations and effectively performing higher-level full BLAS at the innermost loops. Since this is similar in spirit to the vectorization of multifrontal methods, we delay further discussion until Section 6.5.

6.4   Frontal Methods

Frontal methods have their origins in the solution of finite element problems from structural analysis. One of the earliest computer programs implementing the frontal method was that of Irons [100]. He considered only the case of symmetric positive definite systems. The method can, however, be extended to unsymmetric systems [96] and need not be restricted to finite element applications [48]. Indeed, we prefer to view frontal methods as an extension of band or variable-band schemes.

6.4.1   Organization

A common method of organizing the factorization of a band matrix of order n with semibandwidth b is to allocate storage for a full $b \times 2b - 1$ matrix, which we call the frontal matrix, and to use this as a window that runs down the band as the elimination progresses. Thus, at the beginning, the frontal matrix holds rows 1 to b of the band system. This configuration enables the first pivotal step to be performed (including pivoting if this is required); and, if the pivot row is then moved out of the frontal matrix, row $b+1$ of the band matrix can be accommodated in the frontal matrix. One can then perform the second pivotal step within the frontal matrix. Typically, a larger frontal matrix is used since greater efficiency may be possible by moving blocks of rows at a time. It is then usually possible to perform several pivot steps within the one frontal matrix. The traditional reason for this implementation of a banded solver is for the solution of band systems by out-of-core methods since only the frontal matrix need be held in main storage. This use of auxiliary storage is also one of the principal features of a general frontal method.

This "windowing" method can easily be extended to variable-band matrices. In this case, the frontal matrix must have order at least $\max_{a_{ij} \neq 0} \{|i - j|\}$. Further extension to general matrices is possible by observing that any matrix can be viewed as a variable-band matrix. And here lies the main problem with this technique: for any arbitrary matrix with an arbitrary ordering, the required size for the frontal matrix may be very large. However, for discretizations of partial differential equations (whether by finite elements or finite differences), good orderings can usually be found (see, for example, [147] and [64]).

We limit our discussion here to the implementation of Gaussian elimination on the frontal matrix. The frontal matrix is of the form

$$\begin{pmatrix} A & B \\ C & D \end{pmatrix}, \tag{6.5}$$

where A and D are square matrices of order k and r, respectively, where usually $k \ll r, (k+r = m)$. The object at this stage is to perform k steps of Gaussian elimination on the frontal matrix (choosing pivots from A), storing the factors $L_A U_A$ of A, CA^{-1}, and B on auxiliary storage devices, and generating the Schur complement $D - CA^{-1}B$ for use at the next stage of the algorithm. The rows and columns of A are "fully summed"; that is, there will be no further entries in these rows and columns later in the computation. Typically, A might have order 10 to 20, while D is of order 200 to 500.

In the unsymmetric case, the pivots can be chosen from anywhere within A. In our approach [48], we use the standard sparse matrix technique of threshold pivoting, where $a_{ij} \in A$ is suitable as pivot only if

$$a_{ij} \geq u \max(\max_s |a_{sj}|, \max_s |c_{sj}|), \tag{6.6}$$

where u is a preset parameter in the range $0 < u \leq 1$.

Notice that this means that large entries in C can prevent the selection of some pivots from A. Should this be the case, $k_1 \leq k$ steps of Gaussian elimination will be performed, and the resulting Schur complement $D - CA_1^{-1}B$, where A_1 is a square submatrix of A of order k_1, will have order

$r + k - k_1$. Although this can increase the amount of work and storage required by the algorithm, the extra cost is typically very low, and all pivotal steps will eventually be performed since the final frontal matrix has a null **D** block (that is, $r = 0$).

An important aspect of frontal schemes is that all the elimination operations are performed within a full matrix, so that kernels and techniques (including those for exploiting vectorization or parallelism) can be used on full systems. It is also important that k is usually greater than 1, in which case more than one elimination is performed on the frontal matrix.

6.4.2 Vector Performance

We now consider the vectorization of the frontal code MA32 on Cray supercomputers, although the approach is valid for any pipelined vector machine. This work is described in more detail in [24]. The key to optimal performance is to ensure that both the vector floating-point add pipe and the vector floating-point multiply pipes are busy almost all of the time. Specifically, all access to memory must be performed concurrently with the arithmetic. The innermost loop of Gaussian elimination can be written as a SAXPY and therefore can use chaining on the CRAY-1 and CRAY X-MP. On the CRAY-2, the fact that several SAXPYs are being performed allows us to overlap instructions so that the multiply pipe and the add pipe are simultaneously processing different operands. As we shall see shortly, the exploitation of these different features leads to a remarkably similar approach.

```
      DO 200 J = 1,M
         DO 100 I = 1,M
            FA(I,J) = FA(I,J) + PC(I)*PR(J)
100      CONTINUE
200   CONTINUE
```

Figure 6.9: **Inner loops of frontal method**

Figure 6.9 shows the inner loops of a typical frontal method, where the frontal matrix FA is being updated by an outer product between the pivot column PC and the pivot row PR. The code thus represents a rank-one update to the matrix FA. Note that standard loop-unrolling techniques (for example, see Section 1.7) are not possible or, at least, do not give the flexibility we need for high performance. The key to a more efficient implementation is to recognize the nature of the outer-product operations as a single step of Gaussian elimination. If we carry out only one step at a time and assume that the pivot row and column can be held in the registers, the bandwidth of the single port connecting each CPU to the central memory on the CRAY-1 and CRAY-2 computers restricts the performance to half the respective asymptotic rates. However, we can keep the floating-point pipes busy all the time by performing two steps of Gaussian elimination together, using the rank-two update obtained by replacing the calculation in Figure 6.9 by

$$FA(I,J) = FA(I,J) + PC(I,1) * PR(1,J) + PC(I,2) * PR(2,J),$$

where $PC(\cdot,1)$ and $PC(\cdot,2)$ are two pivot columns and $PR(1,\cdot)$ and $PR(2,\cdot)$ are two pivot rows.

The implementation of a rank-two update requires more preparatory work in that two pivots must be chosen and tested for stability. We test the first pivot in the normal way and test the second pivot after updating its column (and row) using the first pivot. This procedure ensures the same robustness as in the original algorithm but does mean some work will have been wasted if the second pivot cannot be chosen from the updated column. We have, however, found this to happen infrequently in practice. Since, for a frontal matrix of order m, all of this preparatory work is $O(m)$ compared to the $O(m^2)$ work of the loop, it will not be very significant when solving large problems.

Of course this rank-two update can be implemented by using a call to the Level 3 BLAS routine GEMM [31] since it is just the update of an m by m matrix FA by adding the product of an m by 2 matrix PC by a 2 by m one PR. However, it is important to notice that almost optimal performance on our target machines can be obtained by only unrolling to this depth. We explain why this is so and give some insight into the tricks necessary for optimal implementation in the following paragraphs.

The implementations of the two pivot schemes on the CRAY-1 and CRAY-2 computers highlight some of the architectural differences between the machines and show how the fine detail of the algorithm design is affected by these differences. In both implementations, the pivot row(s) and column(s) are prefetched from memory. On the CRAY-1, they are held in vector registers, while on the CRAY-2 a combination of local memory and vector registers can be used efficiently. Duff et al. [58] obtained asymptotic rates of 125 Mflops on the CRAY-1 computer using the scheme in Table 6.9, which shows the activity in each of the pipes as time progresses. A common measure of time, which we use here, is a *chime*. For the machines under consideration, this is approximately 70 clocks; its precise value depends on the operation being timed.

L_i and S_i refer, respectively, to the loading or storing of column i of FA, and $P_k R_j (k = 1, 2)$ refers to the triadic vector operation

$$FA(\cdot,J) = FA(\cdot,J) + PC(\cdot,K) * PR(K,J).$$

Table 6.9: **Two-Pivot Algorithm on the CRAY-1**

Pipe	Chime				
	3	4	5	6	7
Load/store	L_2	S_1	L_3	S_2	L_4
$\times/+$ chained	P_2R_1	P_1R_2	P_2R_2	P_1R_3	P_2R_3

Table 6.10: **Two-Pivot Algorithm on the CRAY-2**

Pipe	Chime						
	5	6	7	8	9	10	11
Load/store	L_3	S_1	L_4	S_2	L_5	S_3	L_6
$+$		R_{12}	R_{21}	R_{22}	R_{31}	R_{32}	R_{41}	R_{42}
$\times$		P_{21}	P_{22}	P_{31}	P_{32}	P_{41}	P_{42}	P_{51}

Note that the add and multiply pipes are chained and that there is no chaining of the load pipe although its activity is overlapped with arithmetic on other vectors. We extend this notion of overlap when considering implementation on the CRAY-2.

Since chaining is not available on the CRAY-2, the multiply and add pipes must, in the same chime, be processing separate vectors; that is, we must overlap both floating-point pipes with the load/store pipe. Our implementation on the CRAY-2 can thus be written as in Table 6.10.

Here, $P_{jk}(k = 1, 2)$ corresponds to the operation

$$TEMP_k(\cdot, J) = PC(\cdot, K) * PR(K, J),$$

and R_{jk} corresponds to the operation

$$FA(\cdot, J) = FA(\cdot, J) + TEMP_k(\cdot, J).$$

It is clear that, after the startup, the arithmetic pipes are always busy, so we expect high performance rates for the two-pivot kernel. As in assembler coding for the CRAY-1, instruction timing is critical, and a delay of even a few clocks at the boundaries of the chimes could significantly affect performance. The results shown in Table 6.11 indicate that the present version is not far from optimal.

Of course, developing a fast inner loop is only part of the story. The main concern is the effect of this kernel on the solution of sparse linear equations and, ultimately, on the solution of the

Table 6.11: **Performance of Two-Pivot Kernel on the CRAY-2**

Order of matrix	50	100	500
Megaflops	222	308	385

Table 6.12: **Performance of CRAY-2 Version of MA32 on Element Problems**

	16×16	50×50	100×100
Maximum order of frontal matrix	195	536	1035
Total order of problem	5445	51005	202005
Megaflops	100	302	345

large-scale scientific or engineering problem. We have tested a version of the MA32 package that uses this kernel on the CRAY-2. The problems are artificially generated finite-element problems with rectangular elements having nodes at corners, midpoints of sides, and center and having five variables on each node. We show some results from these runs in Table 6.12, where again high performance is obtained. Note that we are really solving quite large problems; for example, the storage required for the LU factors of the largest problem is over 400 Mwords.

The benefits of vectorization are clear; however, the organization of the frontal method is similar to that for a band matrix solver which is inherently sequential. Thus, we still have the limitation of n sequential pivot steps. This problem is addressed by multifrontal methods.

6.5   Multifrontal Methods

Unless the matrix has a fairly small bandwidth, frontal methods can require more storage and many more floating-point operations than will a more general sparse Gaussian elimination scheme. Multifrontal schemes represent an attempt to retain some of the benefits of frontal schemes while being economic in arithmetic. Multifrontal methods are described in some detail in [56], and their potential for parallelism is discussed in [54, 52, 53].

In this section, we shall work through a small example to give a flavor of the important points and to introduce the notion of an elimination tree. An elimination tree, discussed in detail in [52] and [117], is a tree used to define a precedence order within the factorization. The factorization commences at the leaves of the tree, and data is passed towards the root along the edges in the tree. To complete the work associated with a node all the data must have been obtained from the sons of the node, otherwise work at different nodes is independent. We use the example in Figure 6.10 to illustrate both the multifrontal method and its interpretation in terms of an elimination tree.

$$
\begin{array}{cccc}
\times & \times & \times & \\
& \times & \times & \times \\
\times & \times & \times & \\
\times & \times & & \times \\
\end{array}
$$

Figure 6.10: **Matrix used to illustrate multifrontal scheme**

The matrix shown in Figure 6.10 has a nonzero pattern that is symmetric. (Any system can be considered, however, if we are prepared to store explicit zeros.) The matrix is ordered so that pivots will be chosen down the diagonal in order. At the first step, we can perform the elimination corresponding to the pivot in position $(1,1)$, first "assembling" row and column 1 to get the submatrix shown in Figure 6.11.

$$
\begin{array}{ccc}
\times & \times & \times \\
\times & & \\
\times & & \\
\end{array}
$$

Figure 6.11: **Assembly of first pivot row and column**

By "assembling," we mean placing the nonzeros of row and column 1 into a submatrix of order the number of nonzeros in row and column 1. Thus the zero entries a_{12} and a_{21} have caused row and column 2 to be omitted in Figure 6.11, and so an index vector is required to identify the rows and columns that are in the submatrix. The index vector for the submatrix in Figure 6.11 would have entries $(1,3,4)$ for both the rows and the columns. Column 1 is then eliminated by using pivot $(1,1)$ to give a reduced matrix of order two with associated row (and column) indices 3 and 4. In conventional Gaussian elimination, updating operations of the form

$$a_{ij} = a_{ij} - a_{i1}[a_{11}]^{-1}a_{1j} \qquad (6.7)$$

would be performed immediately for all (i,j) such that $a_{i1}a_{1j} \neq 0$. However, in this formulation, the quantities

$$a_{i1}[a_{11}]^{-1}a_{1j} \qquad (6.8)$$

are held in the reduced submatrix, and the corresponding updating operations are not performed immediately. These updates are not necessary until the corresponding entry is needed in a later pivot row or column. The reduced matrix can be stored until that time.

Row (and column) 2 is now assembled, the (2,2) entry is used as pivot to eliminate column 2, and the reduced matrix of order two—with associated row (and column) indices of 3 and 4—is stored. These submatrices are called frontal matrices. Since more than one frontal matrix generally is stored at any time (currently we have two stored), the method is called "multifrontal." Now, before we can perform the pivot operations using entry (3,3), the updating operations from the first two eliminations (the two stored frontal matrices of order two) must be performed on the original row and column 3. This procedure is effected by summing or assembling the reduced matrices with the original row and column 3, using the index lists to control the summation. Note that this gives rise to an assembled submatrix of order 2 with indices (3,4) for rows and columns. The pivot operation that eliminates column 3 by using pivot (3,3) leaves a reduced matrix of order one with row (and column) index 4. The final step sums this matrix with the (4,4) entry of the original matrix. The sequence of major steps in the elimination is represented by the tree in Figure 6.12.

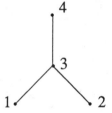

Figure 6.12: **Elimination tree for the matrix of Figure 6.10**

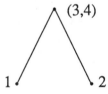

Figure 6.13: **Elimination tree for the matrix of Figure 6.10 after node amalgamation**

The same storage and arithmetic are needed if the (4,4) entry is assembled at the same time as the (3,3) entry, and in this case the two pivotal steps can be performed on the same submatrix. This procedure corresponds to collapsing or amalgamating nodes 3 and 4 in the tree of Figure 6.12 to yield the tree of Figure 6.13. On typical problems, node amalgamation produces a tree with about half as many nodes as the order of the matrix. Duff and Reid [62] employ node amalgamation to enhance the vectorization of a multifrontal approach, and much subsequent work has also used this strategy for better vectorization and exploitation of parallelism (for example, [11] and [117]).

The computation at a node of the tree is simply the assembly of information concerning the node, together with the assembly of the reduced matrices from its sons, followed by some steps of

Gaussian elimination. Each node corresponds to the formation of a frontal matrix of the form (6.5) followed by some elimination steps, after which the Schur complement is passed on for assembly at the father node.

Viewing the factorization as an elimination tree has several benefits. Since only a partial ordering is defined by the tree, the only requirement for a numerical factorization with the same amount of arithmetic is that the calculations must be performed for all the sons of a node before those at the father node can proceed. Thus, many different orderings with the same number of floating-point operations can be generated from the elimination tree. In particular, orderings can be chosen for economic use of storage, for efficiency in out-of-core working, or for parallel implementation (see Section 6.5.2). Additionally, small perturbations to the tree and the number of floating-point operations can accommodate asymmetry, numerical pivoting, or enhanced vectorization.

Liu [117] presents a survey of the role of elimination trees in Gaussian elimination and discusses the efficient generation of trees (in time effectively linear in the number of entries in the original matrix) as well as the effect of different orderings of the tree and manipulations to the tree that preserve properties of the elimination. For example, the ordering of the tree can have a significant effect on the storage required by the intermediate frontal matrices generated during the elimination. Permissible manipulations on the tree include tree rotations, by means of which the tree can, for example, be made more suitable for driving a factorization that exploits parallelism better [146].

This approach can be used on indefinite systems and on matrices that are symmetric in pattern but not in value. However, the only implementations we know of that do this in a stable way are the Harwell codes MA27 (symmetric indefinite systems [61]) and MA37 (symmetric pattern [63]). The MA37 code can be used to solve general systems by using the pattern formed from the Boolean summation of the matrix with its transpose, but the performance is poor on general matrices, as is shown in Table 6.13 from [51].

Table 6.13: **Performance of MA37 on Unsymmetric Patterns on an IBM 3081**

Order	534	1224	183	216
Nonzeros	3474	9613	1069	876
Measure of asymmetry (6.9)	.26	.39	.58	1.0
Nonzeros in factors:				
MA37	9714	91136	2775	7058
MA28	9297	63313	1271	3677
Total storage for subsequent solution				
(in thousands of words):				
MA37	16.3	116.5	5.2	10.6
MA28	21.8	134.0	3.6	8.7

The problems in Table 6.13 are arranged in order of increasing asymmetry, where the measure used is

$$\frac{\text{Number of pairs such that } a_{ij} = 0, a_{ji} \neq 0}{\text{Total number of off-diagonal entries}} \quad (6.9)$$

6.5.1 Performance on Vector Machines

Because multifrontal methods have identical kernels to the frontal methods, one might expect them to perform well on vector machines, although the full matrices involved are usually of smaller dimension and there is a greater amount of data handling outside this kernel. When one is vectorizing multifrontal codes, it is also important to consider not only the kernel but also the assembly operations involved in generating the assembled frontal matrix for the kernel. Amestoy and Duff [5] have used Level 3 BLAS kernels in both assembly and elimination operations to achieve over 225 Mflops on one processor of the CRAY-2 for the factorization of problems from structural analysis. This high computational rate is achieved without incurring a significant increase in arithmetic operations, although the effectiveness of these present multifrontal techniques is limited to matrices whose structure is symmetric (or nearly so). Ashcraft [11] also reports on high computational rates for the vectorization of a multifrontal code for symmetric systems.

6.5.2 Performance on Parallel Machines

An important feature of general elimination trees is that computation at any leaf node can proceed immediately and simultaneously; computations at nodes not on the same direct path from the root to a leaf node are independent. All that is required for computations to proceed at a node is that the calculations at its sons have been completed.

Clearly the parallelism available through the elimination tree depends on the ordering of the matrix. In general, short bushy trees are preferable to tall thin ones since the number of levels determines the inherent sequentiality of the computation. Two common ordering strategies for general sparse symmetric systems are minimum degree and nested dissection (see, for example, [80] and [56]). Although these orderings are similar in behavior for the amount of arithmetic and storage, they give fairly different levels of parallelism when used to construct an elimination tree. We illustrate this point in Table 6.14, where the maximum speedup is computed from a simulation of the elimination as the ratio of the number of operations in the sequential algorithm to the number of sequential operations in the parallel version, with account taken of data movement as well as floating-point calculations [59].

Table 6.14: **Comparison of Two Orderings for Generating an Elimination Tree for Multifrontal Solution** (the problem is generated by a 5-point discretization of a 10×100 grid).

Ordering	Minimum degree	Nested dissection
Number of levels in tree	52	15
Number of pivots on longest path	232	61
Maximum speedup	9	47

If the nodes of the elimination tree are regarded as atomic, then the level of parallelism reduces to one at the root and usually increases only slowly as we progress away from the root. If, however, we recognize that parallelism can be exploited within the calculations at each node (corresponding to one or a few steps of Gaussian elimination on a full submatrix), much greater parallelism can be achieved. In Table 6.15, we give some results of Duff and Johnsson [59] illustrating this effect. Of course, the increase in parallelism comes at the cost of smaller granularity, and the most efficient balance between these opposing effects will depend on the computer architecture.

Table 6.15: **Maximum Speedup Obtainable (minimum degree ordering)**

	30×30 grid	10×100 grid
No parallelism within nodes	3.7	7.5
Parallelism within nodes	30	47

Duff [52] considered the implementation of multifrontal schemes on parallel computers with shared memory. In this implementation, there are really only three types of tasks, the assembly of information from the sons, the selection of pivots, and subsequent eliminations, although we also allow the eliminations to be blocked so that more than one processor could be working on the elimination operations from a single pivot. We choose to store all the tasks available for execution in a single queue with a label to identify the work corresponding to the task. When a processor is free, it then goes to the head of the queue, selects the task there, interprets the label, and performs the appropriate operations. This process may in turn generate other tasks to be added to the end of the queue. We show in Table 6.16 speedup figures that Duff [54] later obtained on the Alliant FX/8 using a modified version of the MA37 code.

Table 6.16: **Speedup on Alliant FX/8 of Five-Point Discretization of Laplace on a** 30×30 **grid**

No. processors	Time	Speedup
1	2.59	-
2	1.36	1.9
4	.74	3.5
6	.57	4.5
8	.46	5.6

6.6 Other Approaches for Exploitation of Parallelism

Although we feel the multifrontal approach to be very suitable for exploitation of parallelism, it is certainly not the only approach being pursued. Indeed, the Cholesky algorithm viewed as a left-looking algorithm (Section 5.4.2) can be implemented for sparse systems and can also be blocked by using a supernodal formulation similar to the node amalgamation that we discussed in Section 6.5. Indeed a code based on this approach attained very high performance on some structural analysis and artificially generated problems on a CRAY Y-MP [146]. A variant of the standard column-oriented sparse Cholesky algorithm has also been implemented on hypercubes ([77] and [78]).

Another approach particularly designed for unsymmetric matrices has been developed by Davis and Yew [25]. Here a set of pivots is chosen simultaneously using Markowitz and threshold criteria so that if the set is permuted to the upper part of the matrix, the corresponding block will be diagonal and all operations corresponding to these pivots can be performed simultaneously. Indeed, it is possible to design an algorithm to perform the pivot search in parallel also. Subsequent sets of independent pivots are chosen in a similar manner until the reduced matrix becomes full enough to switch to full code, as discussed in Section 6.2. Alaghband [2] proposes a similar type of scheme. Other work on the parallel implementation of a multifrontal elimination has been carried out by Benner et al. [15] using large-grain parallelism on a CRAY X-MP and an ELXSI, and by Lucas et al. [118] on a hypercube.

6.7 Software

Although much software is available that implements direct methods for solving sparse linear systems, little is within the public domain. There are several reasons for this situation, the principal ones being that sparse software often is encapsulated within much larger packages (for example, for structural analysis) and that much work on developing sparse codes is funded commercially so that the fruits of this labor often require licenses.

Among the public domain sparse software are some routines from the Collected Algorithms of ACM (available from *netlib*), mostly for matrix manipulation (for example, bandwidth reduction, ordering to block triangular form) rather than for equation solution, although the NSPIV code from Sherman [145] is available as Algorithm 533.

Both Y12M and the Harwell code MA28, referenced in this chapter, are available from *netlib*, although people obtaining MA28 in this way are still required to sign a license agreement. There is also a skeleton sparse LU code from Banks and Smith in the *misc* collection in *netlib*, and Joseph Liu distributes his multiple minimum-degree code upon request.

Among the codes available under license are those from Harwell that were used to illustrate many of the points in this chapter, a subset of which is also marketed by NAG under the title of the Harwell Sparse Matrix Library. Contact details for these organizations can be found in Appendix A.2. IMSL is also developing its own code for solving sparse systems, and a sparse LU code is available in the latest release of ESSL for the IBM 3090. Sparse linear equation codes are also available to users of Cray computers upon request to Cray Research Inc.

Other packages include the SPARSPAK package, primarily developed at the University of Waterloo [82], which solves both linear systems and least-squares problems, and routines in the PORT library from Bell Labs [108], details of which can be obtained from *netlib*. Versions of the package YSMP, developed at Yale University [69], can be obtained from SCA at Yale, who also have several routines implementing iterative methods for sparse equations.

It should be stressed that we have been referring to fully supported products. Many other codes are available that are either at the development stage or are research tools (for example, the SMMS package of Fernando Alvarado at Wisconsin [3]).

6.8   Brief Summary

We have discussed several approaches to the solution of sparse systems of equations, with particular reference to their suitability for the exploitation of vector and parallel architectures. We see considerable promise in both frontal and multifrontal methods on vector machines and reasonable possibilities for exploitation of parallelism by multifrontal methods. A principal factor in attaining high performance is the use of full matrix computational kernels, which have proved extremely effective in the full case.

Chapter 7

Iterative Solution of Sparse Linear Systems

In this chapter we describe a number of the most popular and powerful iterative methods and discuss implementation aspects on vector and parallel computers.

The time-consuming elements (kernels) of iterative methods for solving $Ax = b$ are SAXPYs, SDOTs, and matrix-vector products with A (and often also with a matrix related to A, K^{-1}, the preconditioner). Usually one attempts to optimize each of the specific kernels separately. SAXPYs and SDOTs need little attention, whereas special care is (almost always) required for the matrix-vector products, especially for the preconditioner. Occasionally (see Section 7.1.1, for example), it may be more efficient to combine some of the kernels in order to save on data movement to and from memory.

Iterative methods are in many situations quite attractive for use on vector or parallel computers, especially when the matrix has a regular nonzero structure that can be exploited in obtaining large vector lengths or a structure that leads to (almost) independent substructures. Unfortunately, no single iterative method is robust enough to solve all sparse linear systems accurately and efficiently. Generally, an iterative method is suitable only for a specific class of problem, since the rate of convergence depends strongly on spectral properties of the matrix.

Many, if not most, iterative methods are based upon the following approach. The matrix A is split into an easily invertible part P and a remainder Q. By easily invertible we mean that systems like $Pz = r$ can be solved "cheaply" (P^{-1} is never formed explicitly, unless in trivial situations).

The splitting $A = P - Q$ leads to the basic iteration

$$Px_{i+1} = Qx_i + b, \quad i = 0, 1, 2, \ldots,$$

where x_0 is a user-specified starting vector. This iteration can formally be rewritten as

$$x_{i+1} = P^{-1}(Qx_i + b) = x_i + P^{-1}(b - Ax_i).$$

If we write $r_0 = b - Ax_0$, then it follows by induction that x_i can be expressed as

$$x_i = x_0 + \alpha_0 P^{-1} r_0 + \alpha_1 P^{-1} A P^{-1} r_0 + \ldots + \alpha_i (P^{-1} A)^{(i-1)} P^{-1} r_0.$$

Apparently x_{i+1} is equal to x_0 plus a specific vector from the i-dimensional subspace spanned by the vectors $P^{-1} r_0, P^{-1} A P^{-1} r_0, \ldots, (P^{-1} A)^{(i-1)} P^{-1} r_0$. Such a space of the form

$$\text{span}\{f, Bf, B^2 f, \ldots, B^{(i-1)} r_0\}$$

is called the i-dimensional Krylov subspace corresponding to f and B and is noted as $K_i(B; f)$. In our case we have $x_i = x_0 + y$, with $y \in K_i(P^{-1} A; P^{-1} r_0)$.

When A and P are symmetric positive definite, it is often desirable to work with symmetric operators. To do so, we write $P = LL^T$; by a change of variables, $y = L^T x$. This leads to

$$y_{i+1} = L^{-1} Q L^{-T} y_i + L^{-1} b,$$

and it follows that y_i can be written as $y_i = y_0 + z$, with $z \in K_i(L^{-1} A L^{-T}; L^{-1} r_0)$.

In many applications P is called the preconditioner for the system $Ax = b$. Note that the special choice $P = I$ leads to solution elements belonging to $K_i(A; r_0)$, which gives rise to the unpreconditioned or basic methods.

The straightforward iteration leads to very special elements of the Krylov subspace, but, of course, we are free to select any other member of this subspace. It is then an obvious idea to search for more optimal elements. This approach leads to the so-called Krylov subspace methods or projection-type methods, which include many popular iterative techniques.

Our focus here is on six specific methods:

- CG method. The conjugate gradient method can be used for linear systems of which the matrix is positive definite symmetric.

- LSQR. Of course, the conjugate gradient method can also be applied for general linear systems by applying it to the (not explicitly formed) normal equations. A rather stable way of doing this is achieved by the so-called least squares method. We shall describe a simpler way to obtain a version of the conjugate gradient method that suffers less from rounding errors.

- BG. The biconjugate gradient method can be used for the solution of nonsymmetric linear systems. It has the advantage over the LSQR approach that one avoids forming implicitly the normal equations, so that the corresponding large condition numbers are avoided too. It has the disadvantage that two matrix-vector multiplications are required per iteration step.

- CGS. The so-called conjugate gradient-squared method has become quite popular in recent years. It can be seen as a variant of the BG method insofar as it also exploits the work carried out with matrix-vector multiplications for further improving the convergence behavior.

- GMRES and GMRES(m). For general nonsymmetric systems one can construct a suitable basis for the solution space (which is also done in an elegant way by the previous methods). Unfortunately, this procedure leads to a demand in memory space that grows linearly with the number of iteration steps. The problem is circumvented by restarting the iteration procedure after m steps, which leads to GMRES(m).

- Adaptive Chebychev. If the convex hull of the spectrum of A is known, then the (complex) Chebychev procedure can be used. Of course, such information is only seldom available, but a family of ellipses can be constructed adaptively so that one member of this family makes up the hull, ensuring that the speed of convergence for the Chebychev method is optimal.

Other methods such as SSOR [89], SOR [167], and SIP [153] are not treated here. It will be clear from our discussions, however, that the vectorization and parallelization approaches easily carry over to those methods.

Whatever method is chosen, we suggest monitoring the convergence behavior of an iterative method for a particular problem that (more or less) represents the class of problems one wishes to solve, and comparing this behavior with theoretical convergence results. Such a monitoring process can help identify the effects of rounding errors or reveal programming errors.

7.1 Iterative Methods

7.1.1 Conjugate Gradient

In projection-type methods the idea is to construct a suitable basis for the Krylov subspace

$$K_i(A; r_0) = \text{span}\{r_0, Ar_0, A^2r_0, ..., A^{(i-1)}r_0\}$$

and to solve the system $Ax = b$ projected onto this Krylov subspace.

When A is symmetric, the matrix of the projected system is a symmetric tridiagonal matrix which can be generated by a three-term recurrence relation between the residual vectors. When A is, in addition, positive definite, this tridiagonal system can be solved without any difficulty. This leads to the conjugate gradient algorithm [93, 84].

This method can be described by the following scheme, in which the preconditioner K may be any symmetric positive definite matrix.

$x_0 =$ initial guess; $r_0 = b - Ax_0$;
$p_{-1} = 0; \beta_{-1} = 0$;
solve w_0 from $Kw_0 = r_0$;
$\rho_0 = (r_0, w_0)$
for $i = 0, 1, 2, ...$
    $p_i = w_i + \beta_{i-1}p_{i-1}$;
    $q_i = Ap_i$;
    $\alpha_i = \frac{\rho_i}{(p_i, q_i)}$;
    $x_{i+1} = x_i + \alpha_i p_i$;
    $r_{i+1} = r_i - \alpha_i q_i$;
    if x_{i+1} accurate enough then quit;
    solve w_{i+1} from $Kw_{i+1} = r_{i+1}$;
    $\rho_{i+1} = (r_{i+1}, w_{i+1})$;
    $\beta_i = \frac{\rho_{i+1}}{\rho_i}$;
end.

If we choose $K = I$, then the unpreconditioned, or standard, conjugate gradient method results. For $K \neq I$, we have that, when writing $K = LL^T$, this scheme is equivalent to applying the conjugate gradient method to the (preconditioned) equation $L^{-1}AL^{-T}y = L^{-1}b$, with $x = L^{-T}y$.

Stopping criteria are often based on the norm of the current residual vector r_i. One could terminate the iteration procedure, e.g., when ρ_i/ρ_0 is less than some given value eps. More robust stopping criteria are based on estimating the error in x_i, with respect to x, with information obtained from the iteration parameters α_i and β_i. For details, see [89, 106]. The CG method minimizes the A-norm of the error over the current subspace; in other words, the current iterate x_i is such that $(x_i - x, A(x_i - x))$ is minimal over all x_i in the Krylov subspace spanned by $L^{-1}r_0, L^{-1}r_1, ..., L^{-1}r_{i-1}$. Note that this is accomplished with the aid of only three current vectors r, p, and w. In fact, the CG algorithm can be rewritten as a three-term recurrence relation for the residual vectors r_i, and these vectors form an orthogonal basis for the Krylov subspace.

It can be shown that the number of iteration steps required to get the A-norm of the error below some prescribed eps is roughly proportional to the square root of the condition number of

A (or the preconditioned matrix A, in the preconditioned case). This follows from the well-known upper bound on the residual [12, 84]

$$\|x - x_i\|_A \leq \left(\frac{\sqrt{\kappa} - 1}{\sqrt{\kappa} + 1}\right)^i \|x - x_0\|_A,$$

in which κ is the condition number of $L^{-1}AL^{-T}$.

In practice, the speed of convergence can be considerably faster, most notably in situations in which the extreme eigenvalues are relatively well separated from the rest of the spectrum. The local effects in the convergence pattern of CG have been analyzed in [157].

Since the speed of convergence of the conjugate gradient method depends strongly on the spectrum of the matrix of the system, it is often advantageous to select a "cheap" matrix K (cheap in the sense that $Kw = r$ is easily and efficiently solved) in order to improve the spectral properties. The possibilities in selecting efficient preconditioners K, which allow for a sufficient degree of parallelism and vectorization, will be discussed in Section 7.3, since these aspects are similar for all the methods that follow.

With respect to the different steps in the conjugate gradient method, we make the following remarks.

1. The computation of $p_i = w_i + \beta_{i-1}p_{i-1}$, $q_i = Ap_i$, and (p_i, q_i) can be carried out in combination on successive parts of the vectors. For example, the computation of a segment of the vector q_i can be combined with the contribution from this segment and the corresponding (also available) segment of p_i to the innerproduct. How well the update for p_i can be combined with these two operations depends on the bandwidth of the matrix A. Thus we can exploit the presence of each segment of p_i and q_i in the highest level of memory (vector registers and/or cache memory) by using them more than once. This procedure may help to reduce expensive memory traffic.

2. The same holds for $Kw_{i+1} = r_{i+1}$ and (r_{i+1}, w_{i+1}). The equation $Kw_{i+1} = r_{i+1}$ is usually solved in two steps. With $K = LL^T$, we first solve for z from $Lz = r_{i+1}$, then solve $L^Tw_{i+1} = z$. Note that $(r_{i+1}, w_{i+1}) = (z, z)$, so that the computation of parts of the innerproduct can be combined with solving segments of z from $Lz = r_{i+1}$. For matrices K with a special structure, e.g., block diagonal, these two operations can further be combined with the computation of a section of $r_{i+1} = r_i - \alpha_i q_i$.

3. The performance for the updating of the approximate solutions x_{i+1} can sometimes be improved. Note that the x_i are not necessary for the iteration process itself, and updating

can be carried out only intermittently by storing a number of successive p_i's. Let $P_{i,j}$ denote the matrix with columns $p_i, p_{i+1}, ..., p_{i+j-1}$. If $\alpha_{i,j}$ is defined as the vector with components $\alpha_i, \alpha_{i+1}, ..., \alpha_{i+j-1}$, then the vector x_{i+j} can be formed after j iteration steps as $x_{i+j} = x_i + P_{i,j}\alpha_{i,j}$.

4. The computation of α_i has to be completed before r_{i+1} can be updated. In order to avoid such unwanted synchronization points, alternative conjugate gradient schemes have been proposed, e.g., s-step conjugate gradient [22]. These schemes seem to be more subject to rounding errors [139, 129].

Additional references for CG can be found in [84]. It should also be mentioned that many software libraries contain subroutines for the (preconditioned) CG method, e.g., IMSL, NAG, the Harwell library HSL, and the IBM library ESSL.

7.1.2   Least-Squares Conjugate Gradients

Symmetry of the matrix A (and the preconditioner) is required for the generation of the basis vectors r_i, for the Krylov subspace, by a three-term recurrence relation. Positive definiteness of A is necessary in order to define a suitable norm in which the error has to be minimized. If one of these properties is missing, then the CG algorithm is likely to fail.

A robust but often rather slowly converging iterative method for general linear systems is obtained by applying conjugate gradient to the normal equations for the linear system

$$A^T A x = A^T b.$$

For overdetermined linear systems this leads to the least-squares solution of $Ax = b$. For a discussion of this approach and a comparison with other iterative techniques for overdetermined linear systems, see [158].

For linear systems with a square nonsingular matrix A, the normal equations approach has obvious disadvantages when compared to CG applied to the symmetric positive definite case. First, we have to compute either two matrix-vector products per iteration step (with A and A^T) or only one (with $A^T A$)—but one that is usually less sparse than A, so that the amount of computational work almost doubles. Second, and even more serious, the condition number of $A^T A$ is equal to the square of the condition number of A, so that we may expect a large number of iteration steps.

Nevertheless, in some situations, for example with indefinite systems, the least-squares conjugate gradient approach may be useful. It has been observed that straightforward application of CG to the normal equations may suffer from numerical instability, especially when the matrix of the linear system itself is ill conditioned. According to Paige and Saunders [133] this effect is, to a large extent, due to the explicit computation of the vector $A^T A p_i$. However, the equality $(p_i, A^T A p_i) = (A p_i, A p_i)$ allows us to adjust the scheme slightly. The adjusted scheme, proposed originally by Björck and Elfving [18], is reported to produce better results and runs as follows:

x_0 is an initial guess; $s_0 = b - A x_0$;
$r_0 = A^T s_0; p_0 = r_0$;
$\rho_0 = (r_0, r_0); p_{-1} = 0; \beta_{-1} = 0$;
for $i = 0, 1, 2, \ldots$
$\quad p_i = r_i + \beta_{i-1} p_{i-1}$;
$\quad w_i = A p_i$;
$\quad \alpha_i = \rho_i / w_i, w_i$;
$\quad x_{i+1} = x_i + \alpha_i p_i$;
$\quad s_{i+1} = s_i - \alpha_i w_i$;
$\quad r_{i+1} = A^T s_{i+1}$;
$\quad$ if x_{i+1} is accurate enough then quit;
$\quad \rho_{i+1} = (r_{i+1}, r_{i+1})$;
$\quad \beta_i = \frac{\rho_{i+1}}{\rho_i}$;
end.

Note that r_i is just the residual of the normal equations and $s_i = b - A x_i$ is the residual of the given linear system itself. At each step, the matrices A and A^T are involved; for general sparsity structures, we have to find storage schemes such that both $A p_i$ and $A^T s_{i+1}$ can be computed efficiently.

Another way of reducing the loss of accuracy that accompanies the normal equations approach is to apply the Lanczos algorithm (which is closely related to the conjugate gradient algorithm) to the linear system

$$\begin{pmatrix} I & A \\ A^T & 0 \end{pmatrix} \begin{pmatrix} r \\ x \end{pmatrix} = \begin{pmatrix} b \\ 0 \end{pmatrix}.$$

This forms the basis for the LSQR algorithm proposed by Paige and Saunders [133]. With respect to vector and parallel computing, however, this approach offers no advantages over the Björck and Elfving scheme, and in exact arithmetic it is equivalent to that scheme.

Preconditioning can be applied in three different ways:

1. Solve $P^{-1} A x = P^{-1} b$, which leads to the normal equations

$$A^T P^{-T} P^{-1} A x = A^T P^{-T} P^{-1} b$$

or

$$B^T Bx = B^T P^{-1} b,$$

with

$$B = P^{-1} A.$$

2. Solve $AP^{-1}y = b$, where $x = P^{-1}y$. This leads to solving the normal equations

$$P^{-T} A^T A P^{-1} y = P^{-T} A^T b,$$

or

$$B^T By = B^T b,$$

with

$$B = AP^{-1}.$$

3. With $P = LU$ solve $L^{-1} A U^{-i} y = L^{-1} b$, where $x = U^{-1} y$. The solution y satisfies the normal equations

$$U^{-T} A^T L^{-T} L^{-1} A U^{-1} y = U^{-T} A^T L^{-T} L^{-1} b$$

or

$$B^T By = B^T L^{-1} b,$$

with

$$B = L^{-1} A U^{-1}.$$

In all cases we solve (implicitly) the normal equations for some related linear system; therefore, preconditioning affects the statistical properties of the solution (see, e.g., [158]).

7.1.3   Biconjugate Gradients

For nonsymmetric linear systems it is, in general, not possible to apply the conjugate gradient scheme directly, since orthogonality among the residual vectors r_i cannot be achieved by using simple recurrency relations. Fletcher [71] has proposed an alternative method in which two sequences of mutually orthogonal residuals r_i and $\tilde{r}_i$ are generated by simple relations similar to the conjugate gradient scheme. These r_i and $\tilde{r}_i$ satisfy the relation

$$(r_i, \tilde{r}_j) = 0 \text{ for } i \neq j.$$

The direction vectors p_i and $\tilde{p}_i$ satisfy also a mutual orthogonality condition

$$(\tilde{p}_i, A p_j) = 0, \quad \text{for } i \neq j,$$

which explains the name of the algorithm.

When A has a positive real spectrum and is not defective, a starting residual $\tilde{r}_0$ exists such that the method converges towards the solution of $Ax = b$. Although there is no solid theoretical basis for its convergence behavior, the biconjugate gradient method is efficient for relevant classes of problems. It can be written in the following form (in which K is a suitable preconditioner):

> x_0 is an initial guess; $r_0 = b - Ax_0$;
> solve w_0 from $K w_0 = r_0$;
> $\tilde{r}_0$ is an arbitrary vector such that $(w_0, \tilde{r}_0) \neq 0$,
> usually one chooses $\tilde{r}_0 = r_0$ or $\tilde{r}_0 = w_0$;
> $p_{-1} = \tilde{p}_{-1} = 0; \beta_0 = 0; \rho_0 = (w_0, \tilde{r}_0)$;
> for $i = 0, 1, 2, \ldots$.
> $\quad p_i = w_0 + \beta_i p_{i-1}$;
> $\quad \tilde{p}_i = \tilde{r}_i + \beta_i \tilde{p}_{i-1}$;
> $\quad z_i = A p_i$;
> $\quad \alpha_i = \frac{\rho_i}{(\tilde{p}_i, z_i)}$;
> $\quad r_{i+1} = r_i - \alpha_i z_i$;
> $\quad$ solve $\hat{p}$ from $K^T \hat{p} = \tilde{p}_i$;
> $\quad \tilde{r}_{i+1} = \tilde{r}_i - \alpha_i A^T \hat{p}$;
> $\quad$ solve w_{i+1} from $K w_{i+1} = r_{i+1}$;
> $\quad \rho_{i+1} = (\tilde{r}_{i+1}, w_{i+1})$;
> $\quad$ if $\rho_{i+1} = 0$ then method fails to converge!;
> $\quad x_{i+1} = x_i + \alpha_i p_i$;
> $\quad$ if x_{i+1} is accurate enough then quit;
> $\quad \beta_{i+1} = \frac{\rho_{i+1}}{\rho_i}$;
> end.

With respect to parallel computing, the scheme has the computational advantage over the least-squares approach that the most time-consuming parts $A p_i$ and $A^T \tilde{p}_i$ can be carried out in parallel. When the method is applied to preconditioned linear systems, it is sufficient to construct preconditioners that can be applied in parallel on $p/2$ parallel processors, when p parallel processors are available. An example of such a method is given by Koniges and Anderson [112].

In the biconjugate gradient scheme, there are two pairs of recursions, one pair for r_i and p_i, the other pair for $\tilde{r}_i$ and $\tilde{p}_i$. This latter pair is required only for the construction of the coefficients α_i

and β_i. The fact that, in case of convergence, the vectors r_i and $\tilde{r}_i$ converge to 0 is not exploited by the algorithm.

A possible drawback is that we need to compute Ap_i as well as $A^T\tilde{p}_i$. This may, for matrices with an irregular sparsity structure, lead to storage for both A and A^T if one wishes to optimize both computations. The same remarks hold with respect to the preconditioner K.

7.1.4   Conjugate Gradient Squared

It can easily be verified that the r_i in the unpreconditioned biconjugate gradient scheme (and also in CG) satisfy the relation $r_i = P_i(A)r_0$, in which $P_i(A)$ is an ith degree polynomial in A. The interesting point is that the same polynomial occurs in the expression for $\tilde{r}_i$: $\tilde{r}_i = P_i(A^T)\tilde{r}_0$, and hence

$$\rho_i = (\tilde{r}_i, r_i) = (P_i(A^T)\tilde{r}_0, P_i(A)r_0)$$
$$= (\tilde{r}_0, P_i^2(A)r_0). \tag{7.1}$$

Similar relations hold for the p_i and $\tilde{p}_i$. Apparently the operator $P_i(A)$ transforms the vector r_0 into a small vector r_i, and therefore it might seem a good idea to apply this operator once more, so that in many situations $\hat{r}_i = P_i(A)^2 r_0$ would be doubly reduced.

This was the inspiration to formulate an algorithm in which only the $\hat{r}_i$ are generated (instead of both r_i and $\tilde{r}_i$) [149]. The algorithm was called the "conjugate gradient-squared" method, for obvious reasons. It can be written as the following scheme (in which K is a suitable preconditioner):

x_0 is an initial guess; $r_0 = b - Ax_0$;
$\tilde{r}_0$ is an arbitrary vector, such that
$(r_0, \tilde{r}_0) \neq 0$,
e.g., $\tilde{r}_0 = r_0$; $\rho_0 = (r_0, \tilde{r}_0)$;
$\beta_0 = \rho_0$; $p_{-1} = q_0 = 0$;
for $i = 0, 1, 2, \ldots$
$\quad u_i = r_i + \beta_i q_i$;
$\quad p_i = u_i + \beta_i(q_i + \beta_i p_{i-1})$;
$\quad$solve $\hat{p}$ from $K\hat{p} = p_i$;
$\quad\hat{v} = A\hat{p}$;
$\quad\alpha_i = \frac{\rho_i}{(\tilde{r}_0, \hat{v})}$;
$\quad q_{i+1} = u_i - \alpha_i \hat{v}$;
$\quad$solve $\hat{u}$ from $K\hat{u} = u_i + q_{i+1}$;

$x_{i+1} = x_i + \alpha_i \hat{u}$;
if x_{i+1} is accurate enough then quit;
$r_{i+1} = r_i - \alpha_i A \hat{u}$;
$\rho_{i+1} = (\tilde{r}_0, r_{i+1})$;
if $\rho_{i+1} = 0$ then method fails to converge!;
$\beta_{i+1} = \frac{\rho_{i+1}}{\rho_i}$;
end.

In this scheme the $\hat{r}_i$ have been replaced by r_i. For the conjugate gradient-squared residuals $r(CGS)_i$, when $K = I$, we have that

$$r(CGS)_i = P_i(A)r(BG)_i = P_i(A)^2 r_0, \tag{7.2}$$

in which $r(BG)_i$ is the ith residual in the biconjugate gradient method. As could be expected from this relation, it is often observed in practice that the conjugate gradient-squared method converges roughly twice as fast as the biconjugate gradient method [105, 162].

In spite of these considerations, one should note that a residual vector r_i, in CGS, may have very large components, or more precisely that components of $u_i + q_{i+1}$ take very large values. This situation occurs, for example, when for small components (r_0, v_j) of r_0 in eigendirections v_j the value of $P_i(\lambda_j)$ is large, while $P_i(\lambda_j)(r_0, v_j)$ is still small, so that $P_i(\lambda_j)^2(r_0, v_j)$ may be very large. Usually this effect takes place only during a very short phase of the iteration process, and it can easily be overlooked if one does not check the convergence history. An unpleasant side effect is that these locally large, but successively canceling, corrections degrade the accuracy in the final solution. Hence it may occur that $||r_{i+1}||$, for the recursively updated residual is small, while the actual residual $||Ax_{i+1} - b||$ can be quite large. The usual remedy is to restart the conjugate gradient-squared process as soon as the $||r_i||$ become too large with respect to, say, $||r_0||$.

When the matrix A is symmetric positive definite, BG produces the same x_i and r_i as CG, and hence CGS does not break down.

The situation is less clear when the matrix A is nonsymmetric. A thorough convergence analysis has not yet been given. Surprisingly, however, the method—applied to a suitably preconditioned linear system—often does not suffer from nonsymmetry.

Note that the computational costs per iteration are about the same for BG and CGS, but CGS has the other advantage that only the matrix A itself (and its preconditioner) is involved and not its transpose. For practical implementations in a parallel environment, it might be a slight disadvantage that Ap_i and $A\hat{u}$ cannot be computed in parallel, whereas both the matrix-vector products can in the BG method.

7.1.5   GMRES and GMRES(m)

When A is not symmetric, it is usually impossible to form the projected system (onto the Krylov subspace) by a simple three-term recursion. In fact, the matrix of the projected system is then an upper Hessenberg matrix. Various ways have been proposed to construct such a system. These methods include ORTHOMIN [168], ORTHODIR [101], ORTHORES [101], and GCR [68].

One of the more popular methods in this group is GMRES [140], which is mathematically equivalent to ORTHODIR and GCR, but which has the advantage over ORTHODIR that it does not break down in certain cases where ORTHODIR does. GMRES is also cheaper than GCR both in memory requirements and in arithmetic operations [140].

In GMRES the vector x_i is taken as the vector that minimizes $||b - Ay||_2$ over all y with $y - x_0$ in $K_i(A; r_0)$. Obviously the correction of x_i with respect to x_0 minimizes the residual over this Krylov subspace, so that the residual does not increase when the iteration process proceeds. Since the dimension of the Krylov subspace is bounded by n, the method terminates in at most n steps if rounding errors are absent. In practice, this finite termination property is of no importance, since these iterative methods are attractive only with respect to direct methods (as in Chapter 6) when they deliver a suitable approximation to the solution in far less than n steps. Nevertheless, we may not forget about the effect of rounding errors. They can easily spoil a potentially nice convergence behavior, because of an early loss of orthogonality in the construction of the projected system.

Walker [169] has proposed using Householder transformations in the construction of the Hessenberg matrix, and this approach may be expected to lead to better results in some cases. On the other hand, this method is quite expensive, and in most practical situations the modified Gram-Schmidt orthogonalization method will do almost as well—while being about three times cheaper in arithmetic.

A main practical disadvantage of GMRES (and also of all the other projection-type methods in the nonsymmetric case) is that we have to store all the successive residual vectors and that the construction of the projected system becomes increasingly complex as well. An obvious way to alleviate this disadvantage is to restart the method after each m steps, where m is a suitably chosen integer value. The choice of m requires some skill and experience with the type of problem one wishes to solve. Taking m too small may result in rather poor convergence or even in no convergence at all. In [98], experiments with the choice of m are reported and discussed.

We present the modified Gram-Schmidt version of GMRES(m) for the solution of the linear system $Ax = b$, preconditioned from the left by K^{-1}. In this scheme the required Givens rotations are represented by J_i.

x_0 is an initial guess;
for $j = 1, 2, ...$
    solve r from $Kr = b - Ax_0$;
    $v_1 = r/||r||$;
    $s = ||r||e_1$;
    for $i = 1, 2, ..., m$
        solve w from $Kw = Av_i$;
        for $k = 1, ..., i$ orthogonalization of w
            $h(k, i) = (w, v_k)$; against v's, by modified
            $w = w - h(k, i)v_k$; Gram-Schmidt process
        end k;
        $h(i + 1, i) = ||w||$;
        $v_{i+1} = w/h(i + 1, i)$;
        apply $J_1, ..., J_{i-1}$ on $(h(1, i), ..., h(i + 1, i))$;
        construct J_i, acting on ith and $(i + 1)$st
        component of $h(., i)$, such that $(i + 1)$st component
        of $J_i h(., i)$ is 0;
        $s = J_i s$;
        if $s(i + 1)$ is small enough then (UPDATE$(\tilde{x}, i)$; quit);
        UPDATE$(\tilde{x}, m)$;
    end i;
end j;

In this scheme UPDATE$(\tilde{x}, i)$ replaces the following computations:

Compute y as the solution of $Hy = \tilde{s}$, in which
the upper i by i triangular part of H has $h(i, j)$ as
its elements,
$\tilde{s}$ represents the first i components of s;
$\tilde{x} = x_0 + y_1 * v_1 + y_2 v_2 + \cdots + y_i v_i$;
s_{i+1} equals $||b - A\tilde{x}||$;
if this component is not small enough
then $x_0 = \tilde{x}$;
else quit;

Note that if m has been taken so large that s_{i+1} becomes small enough in the first cycle ($j = 1$), then GMRES(m) is equivalent to GMRES. The GMRES algorithms are more complicated; but for large linear systems and modest values of m, the following components require most of the CPU time:

- the orthogonalization of w against the previous v's,

- the computation of a matrix-vector product with A,

- the solution of a linear system with matrix K, and

- the updating of x_0 to $\tilde{x}$ in the UPDATE step.

These components are essentially the same as for the other iterative methods. They will be discussed in more detail in Sections 7.2 and 7.3.

7.1.6   Adaptive Chebychev

The previously presented methods could be used without any explicit knowledge about the spectrum of the (preconditioned) matrix. However, these methods have the drawback that the iteration constants have to be computed from the iteration vectors. Thus, the computation of these constants serves as a synchronization point, and the continuous computational flow may stagnate. A method that offers better possibilities for parallel implementation is the Chebychev method.

For a successful implementation of this method one needs to have a fairly good estimate for the convex hull of the eigenvalues. This estimate needs to be updated only once per group of iteration steps. The iteration constants can be computed independently of the iteration vectors. In fact, in suitable situations the iteration process can be seen as one continuous flow of computation. The iteration process runs as follows (for given constants d and c, which represent an enclosure of the hull, and for a suitable preconditioner K):

x_0 is an initial guess; $r_0 = b - Ax_0$;
$p_{-1} = 0; \beta_0 = 0; \alpha_0 = 2/d$;
for $i = 0, 1, 2, \ldots$
    solve w from $Kw = r_i$;
    $p_i = w + \beta_i p_{i-1}$;
    $x_{i+1} = x_i + \alpha_i p_i$;
    if x_{i+1} is accurate enough then quit;
    $r_{i+1} = b - Ax_{i+1}(= r_i - \alpha_i Ap_i)$;
    $\beta_{i+1} = (c\frac{\alpha_i}{2})^2$;
    $\alpha_{i+1} = \frac{1}{d-\beta_{i+1}}$;
end.

Note the similarity of this scheme and the conjugate gradient scheme (except for the iteration constants, of course).

The constants d and c should be chosen so that they define a family of ellipses with common center $d > 0$ and foci $d + c$ and $d - c$ (c may be purely imaginary), which contains as a member the ellipse that encloses the spectrum of A and for which the rate of convergence

$$r_c = \frac{a + \sqrt{a^2 - c^2}}{d + \sqrt{d^2 - c^2}}$$

is minimal (a is the length of the x-axis of the ellipse). See for a, d, and c the ellipse in Figure 7.1.

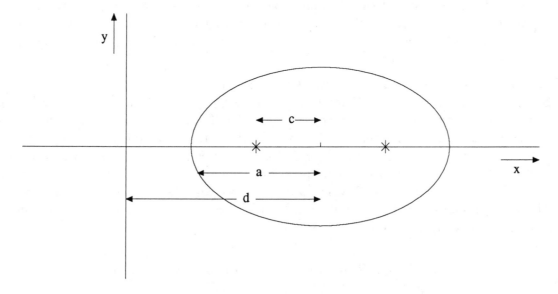

Figure 7.1: **The ellipse that encloses the spectrum of A**

An adaptive procedure to determine d and c such that r_c is close to minimal has been proposed by Manteuffel [120], and a complete Fortran code for the adaptive Chebychev procedure has been published by Ashby [10].

The computational elements for this method are similar to those of CG, except that the inner-products are missing. Hence the method lends itself more to a continuous flow of computation. For example, the computation of segments of p can start right after the first segment of w has emerged; and, similarly, this can be followed directly by the computation of segments of x. Also, segments of r can already be computed before the computation of x has been completed. Finally, the computation of w from $Kw = r$ can start as soon as the elements of r emerge. This procedure can be exploited on some parallel architectures, but it can also be used to improve the performance of the algorithm on vector register computers and on computers with hierarchical memory (e.g., cache). See Section 7.2.2 for an example.

Hybrid algorithms have been proposed (see [84]), in which knowledge about the spectrum of $K^{-1}A$ (obtained by, e.g., CG or GMRES) is used to proceed with the Chebychev method.

7.2 Vector and Parallel Aspects

Usually preconditioning has to be used in combination with the earlier mentioned iterative methods, in order to reduce the number of iteration steps, or sometimes even in order to achieve convergence at all. Popular preconditionings include the so-called incomplete LU decompositions and the incomplete Cholesky decompositions.

In Table 7.1 we show performances in Mflops for a rather large linear system when solved with a straightforward Fortran-coded ICCG method (ICCG = Conjugate Gradients with Incomplete Cholesky preconditioning; for details, see Section 7.3). The linear system was derived by 7-point finite-difference discretization of an elliptic PDE over a 3D rectangular grid, with grid dimensions $n - x = 40, n_y = 40$, and $n_z = 40$, so that the order of the system is $N = 64,000$. Notwithstanding this large value of N, the performances are disappointingly low, and the message is clear. It is highly important to find as much vectorizability and parallelism as possible. Otherwise the applicability of these iterative methods would be very restricted on vector and parallel computers. We address this subject in the following sections.

Table 7.1: **Mflops Rates for the Standard ICCG Method**

Computer	Peak Performance	Straightforward ICCG Method
NEC SX-2	1333	46
CRAY X-MP (2 proc)	470	30
IBM 3090-VF (1 proc)	108	15
CONVEX C-220 (2 proc)	100	6.6
Alliant FX/80 (8 proc)	188	2.6

7.2.1 General Remarks

The basic time-consuming computational elements of the methods presented are as follows:

1. vector updates (SAXPY and GAXPY (generalized SAXPY)),

2. innerproducts (SDOT),

3. matrix-vector products such as Ap (and $A^T p$), and

4. determination of the solution q of systems $Kq = y$, for given y (the preconditioning part of the algorithm), and of systems with K^T (BG and LSQR).

Usually, the SAXPY and SDOT stages can—especially for large systems—be computed efficiently on both vector and parallel computers (or mixed). Also, in many situations the matrix-vector products can be computed with high computational speeds as well, particularly when the matrices have a regular structure.

A common characteristic of these operations is that the amount of data transfer is relatively high with respect to the number of floating-point operations involved. For SAXPY, we observe 2 flops per 3 data transfer operations; for SDOT, the ratio is 2:2; and for Ap, when A has the regular five-diagonal structure and has been stored diagonally, the ratio is 9:11 (a more elaborate analysis is given in [160]).

For most vector computers the optimal ratio is 2:1. In dense matrix computations, such a ratio can often be achieved by rearranging the computations (see, e.g., [34]). With a cache memory, the bandwidth between cache and main memory is a limiting factor; in such a situation one must perform many flops per data transfer between cache and main memory. For dense matrices, these aspects have led to the design of the so-called Level 3 BLAS (see Chapter 5).

In the sparse matrix situation, things are much less rosy. For example, a SAXPY, where the lengths of the vectors exceed the size of the cache, will be executed with a performance significantly below the maximum speed of the computer. This situation can be illustrated by results obtained for the IBM 3090/VF (as well as any other vector machine with a memory hierarchy). Its clock cycle (18.5 nsec) suggests a maximum speed of about 108 Mflops. Since there is only one vector load/store path (as, e.g., with the CRAY-1, Convex, Alliant, and NEC), the performance for DAXPY, when the vectors are available in fastest memory (cache), could be 36 Mflops at most. Because of stripmining effects (see Section 1.2.10) the actual performance is even 20% less: about 29 Mflops (a reduction of 20% from stripmining is also common for other vector computers). However, when the vectors are not already available in cache (which is more or less the standard situation in sparse matrix computations), the performance of DAXPY drops even further, to a modest 19.5 Mflops. Similar performance-reducing effects for SAXPY or DAXPY, as well as for the other computational elements, are observed on most vector computers, and the negative effects in the reported results should in no way be seen as characteristic or special for the IBM 3090/VF.

Occasionally, some of the basic operations in a given iterative method can be grouped together, and then the memory hierarchy can be dealt with more effectively by rearranging the computations. We report on such a situation (again for the IBM 3090/VF) in a subsequent section.

To provide some feeling for actually obtainable computational speeds in a standard situation, we have listed performances for SAXPY and SDOT operations in Table 7.2. All computations have been carried out in about 14 decimal floating-point arithmetic, i.e., single precision for CRAY and double precision for IBM. Since in iterative solution methods the vector lengths are very large (in contrast with direct solution methods), we have listed only the asymptotic performance r_∞.

Table 7.2: **Performances in Mflops for Large Vector Lengths for Some Basic Operations**

Machine	SAXPY	SDOT	Max. Speed
Alliant FX/80 (1 proc.)	3.5	5.1	11.8
Convex C-210	14.8	21	50
IBM 3090/VF	19.5	28	108
CRAY X-MP (1 proc.)	190	206	235
CYBER 205	200	100	200
NEC SX-2	550	905	1333

In a multiprocessor environment, for some of the above machines, performance can be as much as p times larger, where p is the number of processors (see e.g., Table 4.9).

The main bottleneck to achieving reasonable performance is the preconditioning part of the algorithms (or the solution of $Kq = y$), since the most popular choices for K do not immediately lead to vectorizable or parallel code. Therefore we shall discuss this problem in separate sections.

7.2.2   Sparse Matrix-Vector Multiplication

First we shall consider sparse matrix-vector multiplication for the case that A has a very regular structure. Usually the sparse linear systems that arise from finite-difference or finite-element discretizations of partial differential equations have a structure in which the nonzero elements are located only in certain (off-)diagonals of the matrix. As an example we discuss the matrix that results from standard five-point finite-difference approximation of a second-order elliptic PDE over a rectangular grid imposed over a rectangular region in two dimensions:

$$
A = \begin{pmatrix}
\text{x} & \text{x} & & & & & & \text{x} & & & & & & \\
\text{x} & \text{x} & \text{x} & & & & & & \text{x} & & & & & \\
& \text{x} & \text{x} & \text{x} & & & & & & \text{x} & & & & \\
& & \text{x} & \text{x} & \text{x} & & & & & & \text{x} & & & \\
& & & \text{x} & \text{x} & \text{x} & & & & & & \text{x} & & \\
& & & & \text{x} & \text{x} & \text{x} & & & & & & \text{x} & \\
& & & & & \text{x} & \text{x} & & & & & & & \text{x} \\
\text{x} & & & & & & & \text{x} & \text{x} & & & & & \text{x} \\
& \text{x} & & & & & & \text{x} & \text{x} & \text{x} & & & & & \text{x} \\
& & \text{x} & & & & & \text{x} & \text{x} & \text{x} & & & & & & \text{x} \\
& & & & & \cdot & \cdot & \cdot & \cdot & \cdot & & & & \\
& & & & & & \cdot & \cdot & \cdot & \cdot & \cdot & & & \\
& & & & & & \text{x} & & & & & \text{x} & \text{x} & \text{x} \\
& & & & & & & \text{x} & & & & & \text{x} & \text{x} \\
\end{pmatrix}
$$

Figure 7.2: **Matrix that results from standard 5-point finite-difference approximation**

For the derivation of such matrices and their properties, see [167]. When the nonzero entries of the matrix form such a regular systematic pattern, one can choose between storing the nonzeros in a two-dimensional array in row, column, or diagonal fashion. Often the last way is preferred [119], because it leads to larger vector lengths and avoids memory strides. For the problem sketched above, we have given the performances for many supercomputers in the tables in Section 4.4 (loop number 7). We see from these tables that this operation leads to a performance usually in the range of 50–100% of the peak performance of the machine.

For some vector register machines better performance can be obtained when using assembler language. For instance, when we compute $y = Ax$, with A as in Figure 7.2, we need the values $x(i-1)$, $x(i)$, and $x(i+1)$ for the computation of the element $y(i)$. On Cray computers there exists a vector shift instruction that can create the vector streams $x(i+1)$ and $x(i-1)$ from $x(i)$, by shifting the contents of a vector register one position forward or backward; this shift operation can also be chained with other vector operations. Using these techniques, Jordan [104] has shown that this approach can lead to impressive improvements for the CRAY-1.

For computers with a cache memory (and also for some vector register machines) it is sometimes better to exploit the block structure of the matrix. If, in the above situation, the nonzero diagonals of the blocks are stored consecutively, excessive cache refreshment is avoided. However, this depends on the type of computer and the compiler. On most vector register computers the computation of a segment of y is completed before the computation of the next segment of y starts. Thus, the required elements of x for the next segment of y are usually still available in cache.

The performance can be further improved by combining the computation of $y = Ax$ with other suitable parts of the particular method, as, e.g., in CG where $q_i = Ap_i$ can be combined with the computation of the innerproduct (p_i, q_i). Often these situations can be fully exploited in assembler language only, but sometimes the performance may also be improved in standard Fortran.

As an example, we compare two possibilities for the Fortran code for the IBM 3090/VF, and we report on some observed performances. We assume that A is symmetric and that the elements of the upper triangular part of A have been stored by diagonals in 3 arrays. The first possibility is to complete first the computation of $q_i = Ap_i$ in a diagonalwise fashion by code that has as a typical statement

```
      DO 10 J = N1,N2
         Q(J) = A(J-M,3)*P(J-M)+A(J-1,2)*P(J-1)+A(J,1)*P(J)+
     1            A(J,2)*P(J+1)+A(J,3)*P(J+M)
   10 CONTINUE
```

After completion of the matrix-vector product we compute the innerproduct as

```
      PQ = DDOT(N,P,1,Q,1)
```

For matrices of order N=40,000 and bandwidth M=200, the result is a computational speed of 27 Mflops for both elements together. Note that the performance for the innerproduct is negatively influenced by the fact that the elements of P and Q have to be transported all the way through cache again for the computation of PQ.

The next possibility is to combine the two operations. This leads to a code with a typical statement like

```
        DO 10 J = N1,N2
           Q(J) = A(J-M,3)*P(J-M)+A(J-1,2)*P(J-1)+A(J,1)*P(J)+
     1            A(J,2)*P(J+1)+A(J,3)*P(J+M)
           SUM=SUM+Q(J)*P(J)
     10 CONTINUE
```

The computational speed observed for this way of computing is 33 Mflops. Since the performance of the Fortran-coded innerproduct is significantly less than for the assembler-coded DDOT, we believe that, with a careful assembler implementation of the example DO-loop, the performance can be further increased by exploiting the contents of the vector registers and by combining these operations with the computation of p_i (in CG).

Next we consider the case that the nonzero entries are more or less scattered over the matrix A. The storage scheme to be chosen then will depend on the application and on the computer type. A survey of suitable storage schemes for sparse matrices is given in [56], [51], and [136].

As an example, suppose that the matrix has at most five nonzero entries in each row and that one is interested in forming the matrix-vector product. Then the nonzero elements could be stored in an array $AA(N,5)$, whereas the column indices of these elements are supplied in the array $IND(N,5)$. The computation of the required product then reads as follows:

```
        DO 20 J=1,5
           DO 10 I=1,N
              Y(I)=Y(I)+AA(I,J)*X(IND(I,J))
     10     CONTINUE
     20 CONTINUE
```

As a result of the indirect addressing, the performance is much worse than for the regularly structured situation. A loss of at least a factor of 2 in performance has been reported by Dongarra and Duff [33] for many common linear algebra constructions as a result of the use of indirect addressing. Consequently, indirect addressing should be avoided as much as possible.

Also, for more general sparse matrix structures the negative cache effects can be reduced by selecting an appropriate storage scheme. Radicati and Vitaletti [135] compare the performance of the matrix-vector product on the IBM 3090/VF for two different storage schemes. They show that a compressed matrix scheme (as in ESSL [70], ITPACK [110], and ELLPACK [137]) may reduce the CPU time by a factor of over 2 as compared with the more familiar rowwise storage scheme (as in ESSL [70] and also ITPACK [85]).

7.2.3   Performance of the Unpreconditioned Methods

When K in the iterative schemes is chosen to be the identity matrix, then the iterative methods simply consist of SAXPYs, SDOTs, and sparse matrix-vector products. Hence, the performance of the methods is entirely determined by the performance of these computational elements, which have been discussed in some detail before.

Of course, we do not advocate giving up preconditioning, since this may reduce the iteration count significantly. On the other hand, our goal is to reduce CPU time, and therefore performance plays a role. Since preconditioning is in general difficult to vectorize or parallelize, the choice $K = I$ yields the optimal speed for a given method for a particular linear system. This gives us a yardstick with which to measure the performance of implementations of preconditioned iterative methods.

In the most ideal case the computational speed for the preconditioned iteration is close to the speed of the unpreconditioned method (we cannot hope for anything better). This can be far from what we observe in reality. Consider, for example, the results obtained on the CYBER 205 for the discretized Poisson equation with Dirichlet boundary conditions over a unit square with a 30 by 30 rectangular mesh. The matrix of the resulting linear system has the same regular five-diagonal structure as in the previous section. The computational speed for incomplete Cholesky (ICCG), when coded straightforwardly, is then 9.6 Mflops (in vector mode!) on a CYBER 205, whereas the unpreconditioned CG process runs at a speed of 61.6 Mflops. In this case the reduction by a factor of 3 in the number of iteration steps, because of preconditioning, is completely undone by the disappointingly low performance of the preconditioning part of the algorithm. Hence we should, at least in this situation, look for implementations that run at a speed of considerably more than one third of 61.6 Mflops, taking into account the greater computational complexity of the preconditioned algorithm.

All the iterative methods comprise more or less the same building blocks and hence may be expected to run with similar computational speeds. Hence, in Table 7.3, we list the computational speeds for the unpreconditioned CG method only, on different machines, for a linear system with a five-diagonal matrix of dimension 10,000.

Table 7.3: **Computational Speed for the Unpreconditioned CG Process**

Machine	Performance in Mflops	Peak Performance in Mflops	Percentage of Peak Performance
NEC SX-2	643	1333	48
Fujitsu VP-200	300	533	56
Hitachi S810/20	240	800	30
CRAY X-MP (1 proc.)	134	235	57
CYBER 205	106	200	53
CRAY-2 (1 proc.)	82	480	17
IBM 3090-VF (1 proc.)	23	108	21
Convex C-210	19.5	50	39
Alliant FX/4 (1 proc.)	1.8	11	16

The results in Table 7.3 should not be misinterpreted as saying anything about the effectiveness of the CG method. This is largely determined by the number of iteration steps involved. But the results show the limit of performance that can be achieved by any preconditioner. We note that the results have been obtained with straightforward implementations in Fortran 77. In some situations the performance can be significantly higher, when coding in assembler language (CRAY-2) or when coding the algorithm to take advantage of the cache contents (IBM, Alliant).

Note also that all these vector computers behave quite differently for this sparse matrix computation, in that the performance does not simply scale with the peak performance. This is quite unlike some dense matrix computational problems, where—at least for large vector lengths—the peak performance of all these computers can usually be approached.

7.3   Preconditioning

7.3.1   General Aspects

The main difficulty in vectorizing or parallelizing a preconditioned iterative method is usually in the preconditioning part. Since the success of these iterative methods depends (often critically) upon the choice of a suitable preconditioner, we devote much attention to this part of the computation.

First we shall discuss a popular preconditioner for sparse positive definite symmetric matrices, namely, the incomplete Cholesky factorization [126, 127, 84]. Consider the five-diagonal matrix A shown in Figure 7.3:

```
   x                 x       x       x                  x
   x                 x       x       x                  x
 a(i-m,3)        a(i-1,2) a(i,1) a(i,2)              a(i,3)
   x                 x       x       x                  x
   x                 x       x       x                  x
```

Figure 7.3: **Five-diagonal matrix** A

This corresponds to the unknowns over the grid shown in Figure 7.4:

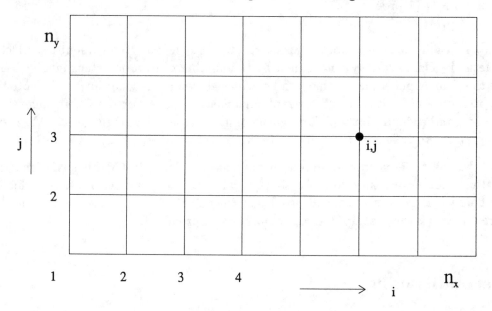

Figure 7.4: **Grid corresponding to the five-diagonal matrix in Fig. 7.3**

The nonzero elements have been represented by $a(i,1)$, $a(i,2)$, and $a(i,3)$. If we write A as $A = L + diag(A) + L^T$, in which L is the strictly lower triangular part of A, then standard incomplete Cholesky (with no fill-in) can be written as

$$K = (L + D)D^{-1}(L^T + D).$$

In this expression K represents the preconditioner to be used in the conjugate gradient scheme (Section 7.1.1). For the standard incomplete Cholesky decomposition the elements $d(i)$ of the diagonal matrix D can be computed from the relation

$$\text{diag}(K) = \text{diag}(A).$$

For the five-diagonal A this leads to the following relations for the $d(i)$:

$$d(i) = a(i,1) - a(i-1,2)^2/d(i-1) - a(i-n_x,3)^2/d(i-n_x).$$

Obviously this is a recursion in both directions over the grid. (This aspect will be discussed later in combination with the application of the preconditioner.)

The so-called modified incomplete decompositions [66, 88] follow from the requirement that rowsum(K) = rowsum(A). This leads to an additional correction to the diagonal $d(i)$. Axelsson and Lindskog [13] describe a relaxed form of this modified incomplete decomposition that, for the five-diagonal A, leads to the following relations for the $d(i)$:

$$\begin{aligned}
d(i) \;=\; & a(i,1) - a(i-1,2)(a(i-1,2) + \alpha a(i-1,3))/d(i-1) \\
& - a(i-n_x,3)(a(i-n_x,3) + \alpha a(i-n_x,2))/d(i-n_x).
\end{aligned}$$

Note that for $\alpha = 0$ we have the standard incomplete Cholesky decomposition described earlier, whereas for $\alpha = 1$ we have the modified incomplete Cholesky decomposition proposed by Gustafsson [88]. It is observed that in many practical situations $\alpha = 1$ does not lead to a reduction in the number of iteration steps, with respect to $\alpha = 0$, but in our experience, taking $\alpha = .95$ almost always reduces the number of iteration steps significantly (by a factor of at least 2). The only difference between the classical and the modified incomplete decomposition is the choice of the diagonal D; the nonzero pattern of the triangular factors is identical (even the off-diagonal elements of the triangular factors are identical).

For the computation of vectors like $w = K^{-1}r$, defined by

$$K^{-1}r = (L^T + D)^{-1}D(L + D)^{-1}r,$$

it will seldom be advantageous to determine the matrices $(L^T + D)^{-1}$ and $(L + D)^{-1}$ explicitly, since these matrices are usually dense triangular matrices.

Instead, for the computation of, say, $y = (L+D)^{-1}r$, y is solved from the linear lower triangular system $(L + D)y = r$. This step then leads typically to relations for the elements $y(i)$, like

$$y(i) = (r(i) - a(i-1,2)y(i-1) - a(i-n_x,3)y(i-n_x))/d(i),$$

which again represents a recursion in both directions over the grid, of the same form as the recursion for the $d(i)$. Note that similar recursions for this nonzero structure of the linear system are encountered when applying iterative methods such as Gauss-Seidel, SOR, SSOR [89], and SIP [153]. Hence these methods can often be made vectorizable or parallel along the same lines as for incomplete Cholesky preconditioning.

Since vector and parallel computers do not lend themselves well to recursions in a straightforward manner, these recursions often degrade the overall performance. Hence, the preceding recursions may seriously degrade the effect of preconditioning on a vector or parallel computer, if carried out in the form given above. These kinds of observation have led to different types of preconditioners, including diagonal scaling, polynomial preconditioning, and truncated Neumann series. Such approaches may be useful in certain circumstances, but they tend to increase the computational complexity (by requiring more iteration steps or by making each iteration step more expensive). On the other hand, various techniques have been proposed to vectorize the recursions, mainly based on reordering the unknowns or changing the order of computation. For regular grids such approaches lead to highly vectorizable code for the standard incomplete factorizations (and consequently also for Gauss-Seidel, SOR, SSOR, and SIP). Before discussing these techniques we shall first present a way to reduce the computational complexity of preconditioning.

7.3.2   Efficient Implementations

Suppose that the given matrix A is written in the form $A = L + \text{diag}(A) + U$, in which L and U are the strictly lower and upper triangular part of A, respectively. Eisenstat [67] has proposed an efficient implementation for the preconditioned iterative methods, when the preconditioner K can be represented as

$$K = (L+D)D^{-1}(D+U), \tag{7.3}$$

in which D is a diagonal matrix. Incomplete Cholesky, incomplete LU, the modified versions of these factorizations, and SSOR can be written in this form. For the incomplete factorizations, we have to ignore all the Gaussian elimination corrections to off-diagonal elements [127]. For the finite-difference discretized operator over rectangular grids, this is equivalent to the standard incomplete factorizations.

The first step to make the preconditioning more efficient is to eliminate the diagonal D in (7.3). We rescale the original linear system $Ax = b$ to

$$D^{-1/2}AD^{-1/2}\tilde{x} = D^{-1/2}b, \tag{7.4}$$

or $\tilde{A}\tilde{x} = \tilde{b}$, with $\tilde{A} = D^{-1/2}AD^{-1/2}$, $\tilde{x} = D^{1/2}x$, and $\tilde{b} = D^{-1/2}b$. With $\tilde{A} = \tilde{L} + \text{diag}(\tilde{A}) + \tilde{U}$, we can easily verify that

(a) $\quad \widetilde{K} = (\tilde{L} + I)(I + \tilde{U})$.

Note that the corresponding triangular systems, like $(\tilde{L}+I)r = w$, are more efficiently solved, since the division by the elements of D is avoided.

We shall further assume that this scaling has been carried out, and the scaled system will again be denoted as $Ax = b$ (note that this scaling does not necessarily have the effect that $\text{diag}(A) = I$).

The key idea in Eisenstat's approach is to apply standard iterative methods (i.e., in their formulation with $K = I$) to the explicitly preconditioned linear system

(b) $\quad (\tilde{L} + I)^{-1}A(I + \tilde{U})^{-1}y = (\tilde{L} + I)^{-1}b$,

in which $y = (I + \tilde{U})x$. This explicitly preconditioned system will be denoted by $Py = c$. Then it follows that

(c) $\quad A = \tilde{L} + I + \text{diag}(A) - 2I + I + \tilde{U}$.

This expression, as well as the special form of the preconditioner given by (a), is used to compute the vector Pz for given z:

$$Pz = (\tilde{L} + I)^{-1}A(I + \tilde{U})^{-1}z = (\tilde{L} + I)^{-1}(z + (\text{diag}(A) - 2I)t) + t, \tag{7.5}$$

with

$$t = (I + \tilde{U})^{-1}z. \tag{7.6}$$

Note that the computation of Pz is equivalent to solving two triangular systems plus the multiplication of a diagonal matrix $(\text{diag}(A) - 2I)$ and a vector and an addition of this result to z. Therefore the matrix-vector product for the preconditioned system can be computed virtually at the cost of the matrix-vector product of the unpreconditioned system. This fact implies that the preconditioned system can be solved by any of the iterative methods for practically the same computational cost per iteration step as the unpreconditioned system. Or, in still other words, the preconditioning comes essentially for free.

In most situations we see, unfortunately, that while we have avoided the fastest part of the iteration process (the matrix-vector product Ap), we are left with the most problematic part of the computation, namely, the triangular solves. But in some circumstances, as we shall see, these parts can also be optimized to about the same level of performance.

7.3.3   Partial Vectorization

A common approach for the vectorization of the preconditioning part of an algorithm is known as *partial vectorization*. In this approach the nonvectorizable loops are split into vectorizable parts and nonvectorizable remainders. Schematically, this approach can be explained as follows.

Assume that the preconditioner is written in the form $K = PQ$, where P is lower triangular and Q is upper triangular. Solving w from $Kw = r$ comes down to solving $Py = r$ and $Qw = y$ successively. Both systems lead to similar vectorization problems, and therefore we consider only the partial vectorization of the computation of y from $Py = r$.

The first step is to regard P as a block matrix with blocks of suitably chosen sizes (not all the blocks need to have equal size):

$$P = \begin{pmatrix} P_{1,1} & & & & \\ P_{2,1} & P_{2,2} & & & \\ P_{3,1} & P_{3,2} & P_{3,3} & & \\ \cdot & \cdot & \cdot & \cdot & \\ \cdot & \cdot & \cdot & \cdot & \\ P_{n,1} & P_{n,2} & \cdot & \cdot & P_{n,n} \end{pmatrix}.$$

Next, we split the vectors y and r in sections $y - i$ and r_i, such that the vector length of the ith section is equal to the block size of $P_{i,i}$. The section y_i is then the solution of

$$P_{i,i}y_i = r_i - (P_{i,1}y_1 + P_{i,2}y_2 + \cdots + P_{i,i-1}y_{i-1}).$$

Note that the amount of work for computing all the sections of y in successive order is equal to the amount of work for solving $Py = r$ in a straightforward manner. However, by rearranging the loops for the subblocks, we see that the right-hand side for each section y_i can be vectorized.

For the five-point finite-difference matrix A, we take the block size equal to n_x, the number of grid points in the x-direction. In that case the incomplete decomposition of A leads to factor P of which the $P_{i,i}$ are lower bidiagonal matrices, the $P_{i,i-1}$ are diagonal matrices, and all of the other $P_{i,j}$ vanish. Hence, the original nonvectorizable three-term recurrence relations are now replaced by two-term recurrence relations of length n_x, and vector statements of length n_x also.

We have thus vectorized half of the work in the preconditioning step, so that the performance of this part almost doubles. In practice the performance is often even better, since for many machines optimized software is available for two-term recurrence relations and this type of computation is often automatically replaced in a Fortran code by the optimized code.

We illustrate the effect of partial vectorization by an example. If our five-diagonal model problem is solved by the preconditioned CG algorithm, the operation count per iteration step is roughly composed as follows: $6N$ flops for the three vector updates, $4N$ flops for the two inner products, $9N$ flops for the matrix-vector product, and $8N$ flops for solving $Kw = r$, if we assume that A has been scaled such that the factors of K have unit diagonal. Assuming that the first $19N$ flops are executed at a very high vector speed and that the preconditioning part is not vectorized and runs at a speed of S Mflops, we conclude, using Amdahl's law (cf. Section 4.1), that the Mflops rate for one preconditioned CG iteration step is approximately given by

$$27/(8/S) \simeq 3.4S \text{ Mflops.}$$

Since for most existing vector computers S is rather modest, the straightforward coded preconditioned CG algorithm (as well as other iterative methods) has a disappointingly low performance. Note that applying Eisenstat's trick (Section 7.3.2) does not lower the CPU time noticeably in this case, since the preconditioning is really the bottleneck.

With partial vectorization we find that the Mflops rate will be approximately

$$27/(4/S_1) \simeq 6.8S_1 \text{ Mflops,}$$

where S_1 is the Mflops rate for a two-term recursion (see loop 5 in Tables 4.2–4.9). For many computers S_1 can be twice as large as S. In practice, the modest blocksize of the subblocks $P_{i,i-1}$ will also often inhibit high Mflops rates for the vectorized part of the preconditioning step. Nevertheless, it is not uncommon to observe that partial vectorization more than doubles the performance.

In Table 7.4 we have listed some performances for the partially vectorized preconditioned CG process (without Eisenstat's trick) for model problems of dimension $N = n_x n_y$, with $n_x=100$, and $n_y=100$. Note that CG involves about $19N$ flops per iteration step and preconditioned CG about $27N$ flops.

Table 7.4: **Mflops Rates for the Partially Vectorized Preconditioned CG and Unpreconditioned CG Process**

Computer	Mflops rate for partially vect. prec. CG	Mflops rate for unpreconditioned CG
NEC SX/2	98	643
CRAY X-MP (8.5 nsec, 1 proc.)	60	134
Hitachi S-810/20	42	238
CRAY-2 (1 proc.)	36	82
Fujitsu VP-200	26	299
IBM 3090/VF	15	23
CYBER 205	12.5	106

From this table we conclude that in most situations the performance of the preconditioned CG process is so low that the reduction in the number of iteration steps (from preconditioning) will often not be reflected by a comparable reduction in CPU time, with respect to the unpreconditioned process. In other words, we have to seek better vectorizable preconditioners in order to beat the unpreconditioned CG process with respect to CPU time.

7.3.4   Reordering the Unknowns

A standard trick is to select all unknowns that have no direct relationship to each other and to number them first. This is repeated for the remaining unknowns. For the five-point finite-difference discretization over rectangular grids, this approach is known as "red-black" ordering. For more complicated discretizations, graph coloring techniques can be used to decouple the unknowns in large groups. In any case the effect is that, after reordering, the matrix is permuted correspondingly and can be written in block form as

$$
A = \begin{pmatrix}
A_{1,1} & A_{1,2} & A_{1,3} & . & . & A_{1,s} \\
A_{2,1} & A_{2,2} & A_{2,3} & . & . & A_{2,s} \\
A_{3,1} & A_{3,2} & A_{3,3} & . & . & A_{3,s} \\
. & . & . & . & . & \\
. & . & . & . & . & \\
A_{s,1} & A_{s,2} & A_{s,3} & . & . & A_{s,s}
\end{pmatrix}
$$

such that all the block matrices $A_{j,j}$ are diagonal matrices. For example, for red-black ordering we have $s = 2$. Then the incomplete LU factorization K of the form

$$K = (L + D)D^{-1}(U + D)$$

with L and U equal to the strict lower and strict upper triangular part of A, respectively, leads to factors $L + D$ and $U + D$ that can be represented by the same nonzero structure, e.g.,

$$L = \begin{pmatrix} D_{1,1} & & & & \\ A_{2,1} & D_{2,2} & & & \\ A_{3,1} & A_{3,2} & D_{3,3} & & \\ \cdot & \cdot & \cdot & \cdot & \\ \cdot & \cdot & \cdot & \cdot & \\ A_{s,1} & A_{s,2} & \cdot & \cdot & D_{s,s} \end{pmatrix}.$$

The corresponding triangular system $(L+D)y = r$ can be solved in an obvious way by exploiting the block structure on a vector or parallel computer. The required matrix-vector products for the subblocks $A_{i,j}$ can be optimized as in Section 7.2.2 (with, of course, smaller vector lengths than for the original system).

For the five-point finite-difference matrix A, the red-black ordering leads to a very vectorizable (and) parallel preconditioner. The performance of the preconditioning step is as high as the performance of the matrix-vector product. This implies that the preconditioned processes, when applying Eisenstat's implementation, can be coded such that an iteration step of the preconditioned algorithm takes approximately the same amount of CPU time as for the unpreconditioned method. Hence any reduction in the number of iteration steps, resulting from the preconditioner, translates immediately to almost the same reduction in CPU time.

One should realize, however, that in general the factors of the incomplete LU factorization of the permuted matrix A are not equal to the similarly permuted incomplete factors of A itself. In other words, changing the order of the unknowns leads in general to a different preconditioner. This fact should not necessarily be a drawback, but often the reordering appears to have a rather strong effect on the number of iterations, so that it can easily happen that the parallelism or vectorizability obtained is effectively degraded by the increase in (iteration) work. Of course, it may also be the other way around—that reordering leads to a decrease in the number of iteration steps as a free bonus to the obtained parallelism or vectorizability.

For standard five-point finite-difference discretizations of second-order elliptic PDEs, Duff and Meurant [60] report on experiments that show that most reordering schemes (including nested dissection and red-black orderings) lead to a considerable increase in iteration steps (and hence in computing time) compared with the standard lexicographical ordering. An exception seems to be a class of parallel orderings introduced in [161], which will be described in Section 7.3.7. As noted before, this may work out differently in other situations, but one should be aware of these possible adverse effects.

Meier and Sameh [124] report on the parallelization of the preconditioned CG algorithm for a multivector processor with a hierarchical memory (e.g., the Alliant FX series). Their approach is based on red-black ordering in combination with forming a reduced system (Schur complement).

7.3.5   Changing the Order of Computation

In some situations it is possible to change the order of the computations (by implicitly reordering the unknowns) without changing the results. This means that bitwise the same results are produced, with the same roundoff effects. The only effect is that the order in which the results are produced may differ from the standard lexicographical ordering. A prime example is the incomplete LU preconditioner for the five-point finite-difference operator over a rectangular grid.

The typical expression in the solution for the lower triangular system is

$$y(i) = (r(i) - a(i-1,2)y(i-1) - a(i-n_x,3)y(i-n_x))/d(i),$$

where $y(i)$ depends only on its previously computed neighbors in the west and south directions over the grid

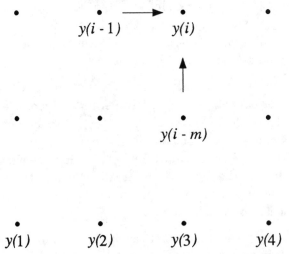

Hence the unknowns $y(i)$ corresponding to the grid points along a diagonal of the grid, i.e., $i+j = constant$, depend only on the values of y corresponding to the previous diagonal. Therefore, if we compute the unknowns in the order corresponding to these diagonals over the grid, then for each diagonal the y-values can be computed independently, or in vector mode.

A number of complications arise, however. If we do not wish to rearrange the unknowns in memory explicitly, then a stride is involved in the vector operations; typically this stride is $n_x - 1$. For some computers a stride is not attractive (e.g., for the CYBER 205 and the Fujitsu VP-200 and for machines with a relatively small cache, such as the IBM 3090/VF), while on others one might encounter a severe degradation in performance because of memory bank conflicts.

The other problem is that the vector length for the preconditioning part is only $\min(n_x, n_y)$ at most, and many of the loops are shorter. Thus, the average vector length may be quite small, and it really depends on the $n_{1/2}$ value for this operation whether the diagonal approach is profitable on a given architecture. Moreover, there are typically more diagonals in the grid than there are grid lines, which means that there are about $n_x + n_y - 1$ vector loops, in contrast to only n_y (unvectorized) loops in the standard lexicographical approach. Some computers have well-optimized code for recursions over one grid line. Again, it then depends on the situation whether the additional overhead for the increased number of loops offsets the advantage of having (relatively short) vector loops.

In Table 7.5 we present some Mflops rates for the preconditioned CG method, applied to the five-diagonal model problem, with the diagonally ordered computation of the preconditioning step. The matrix was obtained from a grid with 200 unknowns in each direction; hence the order is $N = 40,000$.

Table 7.5: **Mflops Rates for the Preconditioned CG Method, with Diagonal Ordering**

Computer	Mflops Rate for Preconditioned CG Method Diagonal Ordering	Mflops Rate for Unpreconditioned CG Procedure
NEC SX/2	482	685
Fujitsu VP-200	208	317
Hitachi S810/20	179	249
CRAY X-MP	118	139

As mentioned before, diagonal ordering of the computation leads to strides in the vectors, which are a problem for some computers. Also, the cache is not optimally exploited in this case. Therefore, we have also coded this approach by explicitly reordering the unknowns corresponding to grid diagonals. To avoid double storage of the matrix coefficients, we have applied Eisenstat's implementation in this case. In Table 7.6 we present some of the results obtained for the same problem as above.

Table 7.6: **Mflops Rates for the Preconditioned CG Method, Applying Diagonal Ordering with Explicitly Storing the Unknowns in Diagonalwise Ordering**

Computer	Mflops Rate for Preconditioned CB Method (Eisenstat impl.)	Mflops Rate for Unpreconditioned CG Procedure
CRAY-2 (1 proc.)	65	76
CYBER 205	56	104
IBM 3090/VF (1 proc.)	16	23

From these tables we conclude that, at least for problems of a reasonable size, applying diagonal ordering to a computation may lead to quite high Mflops rates. These Mflops rates now come rather close to the rates observed for the unpreconditioned process, and we expect them to be approximately the same for very big problems.

For matrices coming from seven-point finite-difference discretizations of elliptic PDEs over rectangular grids in three dimensions, we can vectorize the preconditioning step by combining partial vectorization and diagonalwise ordering. Typically, we compute the solution over planes of the three-dimensional grid, where for each plane we first compute the contribution to the recurrence from the previous plane (partial vectorization in the z-direction). The remaining three-term recursion for the current grid plane is then vectorized by the diagonal ordering approach.

In Table 7.7 we have listed some Mflops rates for the preconditioned CG process, using this approach, for a matrix corresponding to a grid with 40, 39, and 30 unknowns in x, y, z-direction, respectively. Hence the order of the matrix is $N = 46,800$. Note that in this case the vector lengths over the diagonals of each plane are rather limited (less than 40).

Table 7.7: Mflops Rates for the CG Method with Combined Vectorization Techniques for a 3D Problem

Computer	Mflops Rate for Preconditioned CG method	Mflops Rate for Unpreconditioned CG method
NEC SX/2	218	815
Fujitsu VP-200	104	373
Hitachi S-810/20	94	316
CRAY X-MP (1 proc.)	91	152
Convex C-210	15	19
Alliant FX/4 (1 proc.)	1.6	1.8

The short vector lengths in the diagonalwise approach over the planes, in combination with the rather large $n_{1/2}$ values for the Japanese supercomputers, help to explain the modest performance on these machines. For computers with a comparatively small $n_{1/2}$ value, like the Convex, the Alliant, and to a lesser degree the CRAY X-MP, the performances are already close to the performances for the unpreconditioned algorithm.

In three-dimensional problems there are even more possibilities to obtain vectorizable or parallel code. For the standard seven-point finite-difference approximation of elliptic PDEs over a rectangular regular grid, the obvious extension to the diagonal approach in two dimensions is known as the hyperplane ordering. This will be explained in some detail. The typical relation for solving the lower triangular system in three dimensions is as follows:

$$y(i) = (r(i) - a(i-1,2)y(i-1) - a(i-n_x,3)y(i-n_x) - a(i-n,4)y(i-n))/d(i), \qquad (7.7)$$

in which $a(\ ,4)$ represents the elements of A that refer to the connection in the z-direction and $n = n_x \times n_y$. From now on, the unknowns as well as the matrix coefficients will be indicated by three indices i, j, k, so that i refers to the index of the corresponding grid point in the x-direction, and j and k likewise in the y and z-directions. The recursion (7.7) is then rewritten as

$$y(i,j,k) = (r(i,j,k) - a(i-1,j,k,2)y(i-1,j,k) - a(i,j-1,k,3)y(i,j-1,k)$$
$$- a(i,j,k-1,4)y(i,j,k-1))/d(i,j,k). \qquad (7.8)$$

The hyperplane H^m is defined as the collection of grid points for which the triples (i,j,k) have equal sum $i+j+k = m$. Then it is obvious that all unknowns corresponding to H^m can be computed independently (i.e., in vector mode or in parallel) from those corresponding to H^{m-1}. This approach leads to rather long vector lengths $(O(N^{2/3}))$, but the difficulty is that the unknowns required for two successive hyperplanes are not located as vectors in memory, and indirect addressing is the standard technique to identify the unknowns over the hyperplanes.

On some vector supercomputers, indirect addressing can degrade the performance of the computation (e.g., CYBER 205, NEC SX/2); on others, the performance drops to at most about half of the performance that is obtained in the constant stride case (e.g., CRAY X-MP, Fujitsu VP-200, Convex C-210; see also the performance for loop 5 in Tables 4.2–4.9). In [164] the reported performance for ICCG in three dimensions shows that this method, with the hyperplane ordering, can hardly compete with standard conjugate gradient applied to the diagonally scaled system, because of the adverse effects of indirect addressing. However, in [142, 163] ways are presented to circumvent these degradations in performance; for some machines, in particular the CYBER 205 and ETA 10, performances even for the preconditioning part can be obtained that are close to the performances for the regular matrix-vector product. The main idea is to rearrange the unknowns explicitly in memory, corresponding to the hyperplane ordering, where the ordering within each hyperplane is chosen suitably. The more detailed description that follows has been taken from [142].

With respect to the hyperplane ordering, equation (7.8) is replaced by

```
(7.8a)   for m=4,5,6,...,nx+ny+nz
  (a)        y(i,j,k)=r(i,j,k)-a(i,j,k-1,4)*y(i,j,k-1) for (i,j,k) in H{m}
  (b)        y(i,j,k)=y(i,j,k)-a(i,j-1,k,3)*y(i,j-1,k) for (i,j,k) in H{m}
  (c)        y(i,j,k)=y(i,j,k)-a(i-1,j,k,2)*y(i-1,j,k) for (i,j,k) in H{m}
  (d)        y(i,j,k)=y(i,j,k)/d(i,j,k) for (i,j,k) in H{m}
```

We have separated step (d). In practical implementations, it is advisable to scale the given linear system such that $d(i, j, k) = 1$ for all i, j, k. We shall discuss only part (c); the others can be vectorized similarly.

Part (c) is rewritten in the following way:

```
(7.8b)     for i=max(2,m-ny-nz),...,min(nx,m-2)
             for j=max(1,m-i-nz),...,min(ny,m-i-1)
               k=m-i-j
               y(i,j,k)=y(i,j,k)-a(i-1,j,k,2)*y(i-1,j,k)
             end j
           end i   ,
```

This scheme defines the ordering of the unknowns within one hyperplane. The obvious way to implement (7.8b) is to store $y(i, j, k)$ and $a(i-1, j, k, 2)$ in the order in which they are required over H^m. This has to be done only once at the start of the iteration process. Of course, this suggests that we have to store the matrices twice, but this is not really necessary, as we have shown in Section 7.3.2.

Although the elements of y and $a(\ , 2)$ have been reordered, indirect addressing is still required for the elements $y(i-1, j, k)$ corresponding to H^{m-1}. Schematically (7.8b) can be implemented as follows:

1. The required elements $y(i-1, j, k)$ are gathered into an array V, in the order in which they are required to update the $y(i, j, k)$ over H^m. The "gaps" in V, when H^m is larger than H^{m-1}, are left zero.

2. The elements of V are multiplied by the $a(i-1, j, k, 2)$, which are already in the desired order.

3. The result from the previous step is subtracted element-wise from the $y(i, j, k)$ corresponding to H^m.

For the CYBER 205, these three steps can be executed in approximately $2.4 \times n_x \times n_y \times n_z$ clock cycles. Thus, the total time for solving both triangular subsystems is about $14.4N$ clock cycles, where $N = n_x \times n_y \times n_z$. If we compare this with the time required for computing Ay (where A has the nice regular 7-diagonal nonzero structure, scaled so that $\text{diag}(A) = I$), which is about $6 \times N$ clock cycles (see [160]), then we see that there is still a loss by a factor of 2.4 as a result of indirect addressing. For the CYBER 205 and the ETA 10 this loss can be avoided by using sparse vector instructions by which the above operations can be combined into a single vector stream. The procedure leads to a total time for the preconditioning part of about $6N$ clock cycles (see [142, 163]). This is remarkable, since it implies that the incomplete Cholesky preconditioner can be implemented as efficiently as the matrix-vector multiplication.

In order to avoid double storage of A, the matrix-vector multiplication can also be carried out by using the hyperplane ordering. This increases the CPU time only marginally.

On the other hand, since the preconditioner can be executed at virtually the same speed as the matrix-vector product, we can successfully apply the Eisenstat implementation. For a problem with $n_x = 35$, we have observed 91 Mflops for the preconditioned process on the CYBER 205, which is remarkably close to the 104 Mflops for the unpreconditioned process, so that both processes take about as much CPU time per iteration step. (On the other hand, the ETA-10P gives 53 Mflops for the preconditioned process, compared with 90 Mflops for the unpreconditioned method.)

7.3.6   Some Other Vectorizable Preconditioners

Of course, many suggestions have been made for the construction of vectorizable preconditioners. The simplest one is diagonal scaling, by which the matrix A is scaled symmetrically such that the diagonal of the scaled matrix has unit elements. This is known to be quite effective, since it helps to reduce the condition number [73, 156] and often has a beneficial influence on the convergence behavior. On some vector computers the computational speed of the resulting iterative method (without any further preconditioning) is so high that it is often competitive with many of the approaches that have been suggested in previous sections [90, 164].

Nevertheless, in many situations more powerful preconditioners are needed, and many vectorizable variants of these have been proposed. One of the first suggestions was to approximate the inverse of A by a truncated Neumann series [45]. When A is diagonally dominant and scaled such that $\text{diag}(A) = I$, then it can be written as $A = I - B$, and A^{-1} can be evaluated in a Neumann series as

$$A^{-1} = (I - B)^{-1} = I + B + B^2 + B^3 + \cdots. \tag{7.9}$$

Dubois et al. [45] suggest taking a truncated Neumann series as the preconditioner, i.e., approximating K^{-1} as

$$K^{-1} = I + B + B^2 + \cdots + B^p. \tag{7.10}$$

By observing that this preconditioner K^{-1} can be written as a pth degree polynomial $P(A)$ in A, it is obvious that all the iterative methods now lead to iteration vectors x_i in the Krylov subspace that is formed with $P(A)A$ (instead of A, as for the unpreconditioned methods). That is, after m iteration steps we arrive at a Krylov subspace of restricted form with respect to the Krylov subspace that is obtained after $m(p+1)$ iteration steps with the unpreconditioned method. In both cases the amount of work spent in matrix-vector multiplications is the same; hence, at the cost of more iterations, the unpreconditioned process can lead, in theory, to a better approximation for the solution. Therefore it is not plausible that this kind of preconditioning leads to big savings in general.

More sophisticated polynomial preconditioners are obtained when arbitrary coefficients are allowed in the polynomial expansion in A [103, 138]. Apparently they still suffer from the same disadvantage in that they generate approximated solutions in Krylov subspaces of a restricted form, at the cost of the same number of matrix-vector products for which the unpreconditioned method generates a "complete" Krylov subspace. However, they can certainly be of advantage in a parallel environment, since they reduce the effect of synchronization points in the method. Another advantage is that they may lead to an "effective" Krylov subspace in fewer iteration steps with less loss of orthogonality. As far as we know, this point has not yet been investigated.

Obviously, the inverse of A is better approximated by a truncated Neumann series of a fixed degree when A is more diagonally dominant. This is the idea behind a truncated Neumann series approach suggested in [159]. First, an incomplete factorization of A is constructed. To simplify the description, we assume that A has been scaled such that the diagonal elements in the factors of the incomplete decomposition are equal to 1 (see Section 7.3.2):

$$K = (L + I)(I + U). \tag{7.11}$$

Then the factors are written in some suitable block form, e.g.,

$$(L + I) = \begin{pmatrix} L_{1,1} & & & & & \\ L_{2,1} & L_{2,2} & & & & \\ L_{3,1} & L_{3,2} & L_{3,3} & & & \\ . & . & . & . & . & \\ . & . & . & . & . & \\ L_{m,1} & L_{m,2} & . & . & L_{m,m} \end{pmatrix}.$$

If the computation of $L_{i,i}^{-1} r_i$ were an efficient vectorizable operation, then the complete process of solving $(L + I)z = r$ could be vectorized, for segments z_i of z, corresponding to the size of $L_{i,i}$:

$$z = L_{i,i}^{-1}(r_{i,i} - L_{i,i-1}z_i - ... - L_{i,1}z_1) \tag{7.12}$$

(assuming that the operations $L_{i,j}z_j$ are vectorizable operations). In many relevant situations it happens that the factors $L + I$ and $I + U$ are diagonally dominant when A is diagonally dominant, and one might then expect that the subblocks $L_{i,i}$ are even more diagonally dominant. Van der Vorst [159] has proposed using truncated Neumann series only for the inversion of these diagonal blocks of L (and U); he has shown both theoretically and experimentally that only a few terms in the Neumann series, say 2 or 3, suffice to get an efficient (and vectorizable) process, for problems that come from five-point finite-difference approximations in two dimensions.

In most situations there is a price to be paid for this way of vectorization, in that the number of iteration steps increases slightly and also the number of floating-point operations per iteration step increases by $4N$. Van der Vorst [159] has shown for a model problem that the increase in iteration steps is already modest when only 2 terms in the Neumann series are used.

In Table 7.8 the Mflops rates for the 2-term truncated Neumann approach are listed for the 5-point finite-difference matrix in two dimensions. The matrix was obtained from a rectangular grid with 200 unknowns in each direction. Hence the order of the linear system is 40,000.

Table 7.8: **Mflops Rates for CG Method with 2-Term Truncated Neumann Approach for a 2D Problem**

Computer	Preconditioned CG Method	Unpreconditioned CG Method
NEC SX-2	485	685
Fujitsu VP-200	188	317
Hitachi S-810/20	166	249
CRAY X-MP (1 proc.)	125	139
IBM 3090/VF (1 proc.)	19	23
Convex C-210	11.4	11.6
Alliant FX/4 (1 proc.)	1.6	1.9

In [164] this approach is extended to the three-dimensional situation, where $I + L$ can be seen as a nested block form. The resulting method, which has the name "nested truncated Neumann series," leads to rather long vector lengths and can be attractive for some special classes of problem on some vector computers, e.g., the CYBER 205.

Finally, we comment on a vectorizable preconditioner that has been suggested by Meurant [128]. The starting point is a so-called block preconditioner, i.e., a preconditioner of the form

$$K = (L + B)B^{-1}(B + U), \tag{7.13}$$

in which B itself is a block diagonal matrix. This type of preconditioner has been suggested by many authors [23, 125, 109]. Most of these block preconditioners differ in the choice of B. They are reported to be quite effective in two-dimensional situations (in which A is block tridiagonal) in that they significantly reduce the number of iteration steps for many problems. However, in the three-dimensional case, experience leads to less favorable conclusions (see, e.g., [109]). Moreover, for vector computers they share the drawback that the inversion of the diagonal blocks of B (which are commonly tridiagonal matrices) is problematic since it leads to rather low performance.

Meurant [128] has proposed a variant to a block preconditioner introduced in [23], in which he approximates the inverses of these tridiagonal blocks of B by some suitably chosen band matrices. He reports on results for some vector computers (CRAY-1, CRAY X-MP, and CYBER 205) and shows that this approach leads often to lower CPU times than the truncated Neumann series approach.

7.3.7   Parallel Aspects

By reordering the unknowns, a matrix structure can be obtained that allows for parallelism in the triangular factors representing the incomplete decomposition. The red-black ordering, for instance, leads to such a highly parallel form. As has been mentioned before, this reordering often leads to an increase in the number of iteration steps, with respect to the standard lexicographical ordering.

Several attempts have been made to parallelize the standard preconditioned CG process. We shall not comment here on attempts to achieve more parallelism in the conjugate gradient process itself, e.g., by trying to carry out several iteration steps simultaneously, since at the moment these approaches seems to suffer from numerical instability and since they do not carry over easily to the other methods. Of more interest is the effort that has been put into constructing a parallel preconditioner, since these attempts are also relevant for the other iterative methods.

Let us write the triangular factors of K in block bidiagonal form:

$$
L = \begin{pmatrix}
L_{1,1} & & & & & \\
L_{2,1} & L_{2,2} & & & & \\
& L_{3,2} & L_{3,3} & & & \\
& & \cdot & \cdot & \cdot & \\
& & & \cdot & \cdot & \cdot \\
& & & & L_{p,p-1} & L_{p,p}
\end{pmatrix}
$$

For p not too small, Seager [144] suggests setting some of the off-diagonal blocks of L to zero (and to do so in a symmetrical way in the upper triangular factor U, too). Then the back substitution process is decoupled into a set of independent back substitution processes. The main disadvantage of this approach is that often the number of iteration steps increases, especially when more off-diagonal blocks are discarded.

Radicati and Robert [134] suggest constructing in parallel incomplete factorizations of slightly overlapping parts of the matrix. They report that this sometimes leads to a decrease in the number of iteration steps (in experiments carried out on a six-processor IBM 3090/VF).

Still another approach, suggested by Meurant [129], exploits the idea of the two-sided (or twisted) Gaussian elimination procedure for tridiagonal matrices [14, 161]. This is generalized for the incomplete factorization. In this approach K is written as PQ, where P takes the (twisted) form

$$\begin{pmatrix}
P_{1,1} & & & & & & & \\
P_{2,1} & P_{2,2} & & & & & & \\
& P_{3,2} & P_{3,3} & & & & & \\
& & & \cdot & \cdot & & & \\
& & & P_{p-1,p} & P_{p,p} & P_{p,p+1} & & \\
& & & & & P_{p+1,p+1} & P_{p+1,p+2} & \\
& & & & & & \cdot & \cdot \\
& & & & & & & P_{q,q}
\end{pmatrix}$$

(and Q has a block structure similar to P^T). This approach can be viewed as starting the (incomplete) factorization process simultaneously at both ends of the matrix A.

Van der Vorst [161] has shown how this procedure can be done in a nested way for the diagonal blocks of P (and Q) too. For the two-dimensional five-point finite-difference discretization over a rectangular grid, the first approach comes down to reordering the unknowns (and the corresponding equations) as

```
   1   3   5   7   9  11
  13  15  17  19  21  23
  --->

           .   .   .   .   .

  --->
  14  16  18  20  22  24
   2   4   6   8  10  12      ,
```

while the nested (twisted) approach is equivalent to reordering the unknowns as

```
   1   5   9  13     14  10   6   2
  17  21  25  29     30  26  22  18
  33  37  41  45     46  42  38  34

  35  39  43  47     48  44  40  36
  19  23  27  31     32  28  24  20
   3   7  11  15     16  12   8   4
```

That is, we start to number from the four corners of the grid in an alternating manner. It is obvious that the original twisted approach leads to a process that can be carried out almost entirely in two parallel parts, while the nested form can be done almost entirely in four parallel parts. Similarly, in three dimensions the incomplete decomposition, as well as the triangular solves, can be done almost entirely in eight parallel parts.

Van der Vorst [161, 163] reports a slight decrease in the number of iteration steps for these parallel versions, with respect to the lexicographical ordering. Duff and Meurant [60] have compared the preconditioned conjugate gradient method for a large number of different orderings, such as nested dissection and red-black, zebra, lexicographical, Union Jack, and nested parallel orderings. The nested parallel orderings are among the most efficient; thus, they are good candidates even for serial computing, and parallelism here comes as a free bonus.

At first sight there might be some problems for parallel vector processors, since in the sketched orderings the subsystems are in lexicographical ordering and hence not completely vectorizable. Of course, these subgroups could be reordered diagonally:

$$
\begin{array}{ccccccccc}
1 & 5 & 13 & . & . & . & 14 & 6 & 2 \\
9 & 17 & . & . & . & . & . & 18 & 10 \\
21 & . & . & . & . & . & . & & 22 \\
. & . & . & . & . & . & . & . & . \\
23 & . & . & . & . & . & . & . & 24 \\
11 & 19 & . & . & . & . & . & 20 & 12 \\
3 & 7 & 15 & . & . & . & 16 & 8 & 4 \quad ,
\end{array}
$$

which then leads to vector code as shown in Section 7.3.5. The disadvantage is that in practical situations the vector lengths will only be small on average. In [163], alternative orderings are suggested, based on carrying out the twisted factorization in a diagonal fashion. For example, in the two-dimensional situation the ordering could be

$$
\begin{array}{ccccccccc}
1 & 3 & 7 & 13 & . & . & . & . & . \\
5 & 9 & 15 & . & . & . & . & . & . \\
11 & 17 & . & . & . & . & . & . & . \\
19 & . & . & . & . & . & . & . & . \\
. & . & . & . & . & . & . & . & 20 \\
. & . & . & . & . & . & . & 18 & 12 \\
. & . & . & . & . & . & 16 & 10 & 6 \\
. & . & . & . & . & 14 & 8 & 4 & 2 \quad ,
\end{array}
$$

which leads to a process that can be done almost entirely in parallel (except for the grid diagonal in the middle, which is coupled to both groups), and each group can be done in vector mode, just as shown in Section 7.3.5. Of course, this can be generalized to three dimensions, leading to four parallel processes, each vectorizable. The twisted factorization approach can also be combined with the hyperplane approach in Section 7.3.5.

In [161] it is mentioned that these twisted incomplete factorizations can be implemented in the efficient manner proposed by Eisenstat [67] (see also Section 7.3.2), since they satisfy the requirement that the entries in corresponding locations in A be equal to the off-diagonal elements of $\tilde{P}$ and $\tilde{Q}$ in

$$K = ((\tilde{P} + D)D^{-1})(D + \tilde{Q}), \tag{7.14}$$

with $\tilde{P} + D = PD^{-1/2}$, $D^{-1/2}Q = D + \tilde{Q}$.

7.4   Experiences with Parallelism

7.4.1   General Remarks

Although the problem of finding efficient parallel preconditioners has not yet been solved, it may be helpful to present some experimental results for some of the previously discussed approaches. These results should be considered as representing the state of the art rather than the ultimate answer to our parallel demands and possibilities. All of the results have been reported for the nicely structured systems coming from finite-difference discretizations of elliptic PDEs over two-dimensional and three-dimensional rectangular grids.

7.4.2   Overlapping Local Preconditioners

Radicati and Robert [134] suggest partitioning the given matrix A in (slightly) overlapping blocks along the main diagonal, as in Figure 7.5.

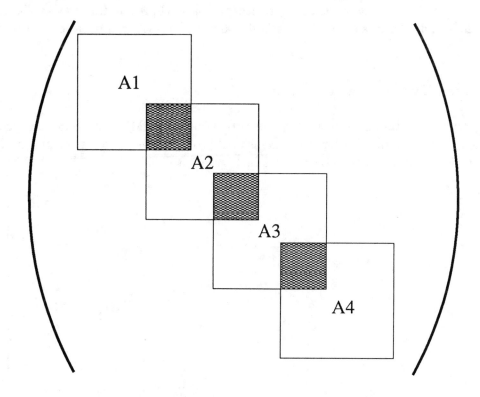

Figure 7.5: **Overlapping blocks in A**

Note that a given nonzero entry of A is not necessarily contained in one of these blocks. But experience suggests that this approach is more successful if these blocks cover all the nonzero entries of A. The idea is to compute in parallel local preconditioners for all of the blocks, e.g.,

$$A_n = L_n D_n^{-1} U_n - R_n. \tag{7.15}$$

Then, when solving $Kw = r$ in the preconditioning step, we partition r in (overlapping) parts r_n, according to A_n, and we solve the systems $L_n D_n^{-1} U_n w_n = r_n$ in parallel. Finally we define the elements of w to be equal to corresponding elements of the w_n's in the nonoverlapping parts and to the average of them in the overlapped parts.

Radicati and Robert [134] report on timing results obtained on an IBM 3090-600E/VF for GMRES preconditioned by overlapped incomplete LU decomposition for a two-dimensional system of order 32,400 with a bandwidth of 360. For p processors ($1 \leq p \leq 6$) they subdivide A in p overlapping parts, the overlap being so large that these blocks cover all the nonzero entries of A.

They found experimentally an overlap of about 360 elements to be optimal for their problem. This approach led to a speedup of roughly p. In some cases the parallel preconditioner was slightly more effective than the standard one, so that this method is also of interest for applications on traditional computers.

7.4.3   Repeated Twisted Factorization

Meurant [130] reports on timing results obtained with a CRAY Y-MP/832, using an incomplete repeated twisted block factorization for two-dimensional problems. In his experiments the L of the incomplete factorization has a block structure as in Figure 7.6.

$$
L = \begin{pmatrix}
X & & & & & & & & & & & & & & & & & & \\
X & X & & & & & & & & & & & & & & & & & \\
  & X & X & & & & & & & & & & & & & & & & \\
  & & & X & X & & & & & & & & & & & & & & \\
  & & & & X & X & & & & & & & & & & & & & \\
  & & & & & X & & & & & & & & & & & & & \\
  & & & & & & X & & & & & & & & & & & & \\
  & & & & & & X & X & & & & & & & & & & & \\
  & & & & & & & X & X & & & & & & & & & & \\
  & & & & & & & & & X & X & & & & & & & & \\
  & & & & & & & & & & X & X & & & & & & & \\
  & & & & & & & & & & & X & & & & & & & \\
  & & & & & & & & & & & & X & & & & & & \\
  & & & & & & & & & & & & X & X & & & & & \\
  & & & & & & & & & & & & & X & X & & & & \\
  & & & & & & & & & & & & & & & X & X & & \\
  & & & & & & & & & & & & & & & & X & X & \\
  & & & & & & & & & & & & & & & & & X &
\end{pmatrix}
$$

Figure 7.6: **Block structure of left decomposition factor L**

Specifically, L has alternatingly a lower block diagonal, an upper one, a lower one, and (finally) an upper one. For this approach Meurant reports a speedup, for preconditioned CG, close to 6 on the 8-processor CRAY Y-MP. This speedup has been measured relative to the same repeated twisted factorization process executed on a single processor. Meurant also reports an increase in the number of iteration steps as a result of this repeated twisting. This increase implies that the effective speedup with respect to the nonparallel code is only about 4. As far as we know, Meurant's approach has not been combined with the approach of Radicati and Robert.

7.4.4 Twisted and Nested Twisted Factorization

For three-dimensional problems on a two-processor system we have used the blockwise twisted approach in the z-direction. That is, the (x, y)-planes in the grid were treated in parallel from bottom and top inwards. Over each plane we used diagonalwise ordering, in order to achieve high vector speeds on each processor.

On a CRAY X-MP/2 this led, for a preconditioned CG, to a reduction by a factor of close to 2 in wall clock time with respect to the CPU time for the nonparallel code on a single processor. For the microtasked code the wall clock time on the 2-processor system was measured for a dedicated system, whereas for the nonparallel code the CPU time was measured on a moderately loaded system. In some situations we even observed a reduction in wall clock time by a factor of slightly more than two, because of better convergence properties of the twisted incomplete preconditioner.

As suggested before, we can apply the twisted incomplete factorization in a nested way. For three-dimensional problems this can be exploited by twisting also the blocks corresponding to (x, y) planes in the y-direction. Over the resulting blocks, corresponding to half (x, y) planes, we may apply diagonal ordering in order to fully vectorize the four parallel parts. With this approach we have been able to reduce the wall clock time by a factor of 3.3, for a preconditioned CG, on the 4-processor Convex C-240. In this case the total CPU time, used by all of the processors, is roughly equal to the CPU time required for single-processor execution.

Other than for the experiments on the CRAY X-MP/2 reported earlier, we have relied on the autotasking capabilities of the Fortran compiler for the Convex C-240 for all of the code except the preconditioning part. One reason that the factor 3.3 for the Convex C-240 stays well behind the theoretically expected factor of about 3.9 may be that some statements in the code lead to rather short vector lengths. Another reason might be that we were not completely sure whether our testing machine was executing constantly in standalone mode during the time of our timing experiments. For more details on these and other experiments, see [165].

7.4.5 Hyperplane Ordering

As has been shown in Section 7.3.5, the hyperplane ordering can be used to realize long vector lengths in three-dimensional situations—at the expense, however, of indirect addressing.

For a CYBER 205, we have demonstrated in some detail how these adverse indirect addressing effects can be circumvented. A similar approach has been followed by Berryman et al. [16] for parallelizing standard ICCG on a Connection Machine Model 2. For a 4K-processor machine they report a computational speed of 52.6 Mflops for the (sparse) matrix-vector product, while 13.1 Mflops has been realized for the preconditioner with the hyperplane approach. This reduction in speed by a factor of 4 makes it attractive to use only diagonal scaling as a preconditioner, in certain situations, for massively parallel machines like the CM-2. The latter approach has been followed by Mathur and Johnsson [122] for finite-element problems.

We have used the hyperplane ordering for preconditioned CG on an Alliant FX/4, for three-dimensional systems with dimensions $n_x = 40, n_y = 39$, and $n_z = 30$. For 4 processors this led to a speedup of 2.61, compared with a speedup of 2.54 for the CG process with only diagonally scaling as a preconditioner. The fact that both speedups are far below the optimal value of 4 must be attributed to cache effects. These cache effects can be largely removed by using the reduced system approach suggested by Meier and Sameh [124]. However, for the three-dimensional systems that we have tested, the reduced system approach led, on the average, to about the same CPU times as for the hyperplane approach on Alliant FX/8 and FX/80 computers.

Appendix A

Acquiring Mathematical Software

A1. Netlib

We have collected in *netlib* much of the software described or used in this book. The *netlib* service provides quick, easy, and efficient distribution of public-domain software to the scientific computing community on an as-needed basis.

A user sends a request by electronic mail to *netlib@ornl.gov* on the Internet. A request has one of the following forms:

```
send index

send index from {library}

send {routines} from {library}

find {keywords}
```

The *netlib* service provides its users with features not previously available:

- There are no administrative channels to go through.

- Since no human processes the request, it is possible to get software at any time, even in the middle of the night.

- The most up-to-date version is always available.

191

• Individual routines or pieces of a package can be obtained instead of a whole collection.

Below is a list of software available when this book went to print.

```
This directory contains software described in the book,
Linear Algebra Computations on Vector and Parallel Computers,
by Jack Dongarra, Iain Duff, Danny Sorensen, and Henk Van der Vorst.

slpsubhalf  software to measure the performance of routines SGEFA and
            SGESL from the Linpack package.

linpacks    software to run the "Linpack Benchmark"

sblas1 single precision Level 1 BLAS
dblas1 double precision Level 1 BLAS
cblas1 complex precision Level 1 BLAS
zblas1 double complex precision Level 1 BLAS

sblas2 single precision Level 2 BLAS
dblas2 double precision Level 2 BLAS
cblas2 complex precision Level 2 BLAS
zblas2 double complex precision Level 2 BLAS

sblas3 single precision Level 3 BLAS
dblas3 double precision Level 3 BLAS
cblas3 complex precision Level 3 BLAS
zblas3 double complex precision Level 3 BLAS

slus.f single precision versions of different blocked
       LU decomposition algorithms
schol.f single precision versions of different blocked
        Cholesky decomposition algorithms
sqrs.f single precision versions of different blocked
       QR decomposition algorithms

benchm a benchmark program for performance test for Fortran loops.
       The program executes a number of Fortran DO-loops and lists
       the execution times for different loop lengths, the Mflops-rates
       and the performance parameters R-inf and n-half.
```

      `pcg3d a preconditioned conjugate gradient code for 3D problems.`

To get an up-to-date listing, send electronic mail to

    `netlib@ornl.gov`

In the mail message, type

    `send index from DDSV`

A1.1 Mathematical Software

In addition netlib contains various items of software and test problems which maybe useful. Listed below are some of these items. (To obtain an item type: *send ma28ad from harwell*, for example.)

```
(harwell/ma28ad) lu,general sparse matrix,pivot for sparsity and stability
(harwell/ma28bd) lu,sparse,different values,factored by (harwell/ma28ad)
(hompack/fixpds) nonlinear equations,f(x)=0,sparse,ode based
(hompack/fixpds) nonlinear equations,rho(a,lambda,x)=0,sparse,ode based
(hompack/fixpds) nonlinear equations,x=f(x),sparse,ode based
(hompack/fixpns) nonlinear equations,f(x)=0,sparse,normal flow
(hompack/fixpns) nonlinear equations,rho(a,lambda,x)=0,sparse,normal flow
(hompack/fixpns) nonlinear equations,x=f(x),sparse,normal flow
(hompack/fixpqs) nonlinear equations,f(x)=0,sparse,augmented jacobian
(hompack/fixpqs) nonlinear equations,rho(a,lamb,x)=0,sparse,augmented jacobian
(hompack/fixpqs) nonlinear equations,x=f(x),sparse,augmented jacobian
(laso/dilaso) all eigenvalue and eigenvector,sparse symmetric,lanczos
(laso/dnlaso) few eigenvalue and eigenvector,sparse symmetric,lanczos
(misc/bsmp) bank and smiths sparse lu made simple
(odepack/lsodes) stiff and nostiff odes,sparse jacobian
(port/chk/prac) test of the port sparse matrix package,[cp]
(port/chk/prad) test of the port sparse matrix package,[dp]
```

```
(port/chk/prsa) test of the port sparse matrix package,[sp]
(port/ex/prea) example of (port/spfor),row and column ordering,sparse matrix
(port/ex/prma) example of (port/spmsf),symbolic lu decomposition,sparse matrix
(port/ex/prs1) example of (port/spmml),sparse matrix vector multiplication
(port/ex/prs3) example of (port/spmnf),numerical lu decomposition,sparse matrix
(port/ex/prsa) example of (port/spmle),sparse linear system solution
(port/ex/prsf) example of (port/spfle),sparse linear system solution
(port/ex/prsj) example of (port/spmce),lu decomposition,sparse matrix
(port/ex/prsm) example of (port/spfce),lu decomposition,sparse matrix
(port/ex/prsp) example of (port/spmlu),lu decomposition of a sparse matrix
(port/ex/prst) example of (port/spflu),lu decomposition of a sparse matrix
(port/ex/prsy) example of (port/spfnf),lu decomposition,sparse matrix
(port/ex/prsz) example of (port/spfml),sparse matrix vector multiplication
(toms/508) bandwidth reduction,profile reduction,sparse matrix
(toms/509) bandwidth reduction,king algorithm,profile reduction,sparse matrix
(toms/529) symmetric permutations,block triangular,depth first search,sparse
(toms/533) sparse matrix,simultaneous linear equations,partial pivoting
(toms/538) eigenvalue,eigenvector,sparse,diagonalizable,simultaneous iteration
(toms/570) eigenvalue,eigenvector,iteration,real sparse nonsymmetric matrix
(toms/586) iterative methods,sparse matrix
(toms/601) sparse matrix
(toms/618) estimating sparse jacobian matrices
(toms/636) estimating sparse hessian matrices,difference of gradients
(y12m/y12m) solution of linear systems,matrices are large and sparse
(y12m/y12mae) sparse matrix,solve one system,single right hand side,[sp]
(y12m/y12maf) sparse matrix,solve one system,single right hand side,[dp]
```

A2. Mathematical Software Libraries

Large mathematical libraries of mathematical software are maintained by the International Mathematical and Statistical Libraries (better known as IMSL), the Numerical Algorithms Group (better known as NAG), and the Harwell Subroutine Library. IMSL and NAG provide their libraries under a license agreement with support provided. With the Harwell Subroutine Library license agreement, no support is provided.

IMSL, Inc.
2500 Park West Tower 1
2500 City Blvd.
Houston, Texas 77042-3020
713-782-6060

NAG, Inc.
1400 Opus Place, Suite 200
Downers Grove, IL 60515-5702
708-971-2337, FAX 708-971-2706

NAG Ltd.
Wilkinson House
Jordan Hill Road
Oxford OX8 YDE
England
+44-865-511245, FAX +44-865-310139

Harwell Subroutine Library
Mr. S. Marlow
Building 8.9
Harwell Laboratory
Didcot, Oxon, OX11 0RA, England

MathWorks

In addition to the Fortran mathematical software libraries, MathWorks provides an interactive system for matrix computations called MATLAB. This matrix calculator runs on many systems from personal computers to workstations to mainframes. MATLAB provides a programming language to allow rapid algorithm design which facilitates the construction and testing of ideas and algorithms. For information on the license agreement for MATLAB, contact MathWorks:

The MathWorks, Inc.
20 North Main St.
Sherborn, MA 01770
508-653-1415

Appendix B

Glossary

address generation: During the execution of an instruction, the cycle in which an effective address is calculated by means of indexing or indirect addressing.

Amdahl's law: The relationship between performance (computational speed) and CPU time. When two parts of a job are executed at a different speed, the total CPU time can be expressed as a function of these speeds. Amdahl first pointed out that the lower of these speeds may have a dramatic influence on the overall performance.

array constant: Within a DO-loop, an array reference all of whose subscripts are invariant:

```
    DO 10 I = 1, N
        A(I) = X(J) * B (I+J) + Z(8,J,K,3)
 10 CONTINUE
```

In the preceding loop, X(J) and Z(8,J,K,3) are array constants. (A(I) and B(I+J) are not array constants since the loop index appears in the subscripts.)

associative access: A method in which access is made to an item whose key matches an access key, as distinct from an access to an item at a specific address in memory.

associative memory: Memory whose contents are accessed by key rather than by address.

attached vector-processor: A specialized processor for vector computations, designed to be connected to a general-purpose host processor. The host processor supplies input/output functions, a file system, and other aspects of a computing system environment.

automatic vectorization: A compiler that takes code written in a serial language (usually Fortran or C) and translates it into vector instructions. The vector instructions may be machine specific or in a source form such as array extensions or as subroutine calls to a vector library.

auxiliary memory: Memory that is usually large, slow, and inexpensive, often a rotating magnetic or optical memory, whose main function is to store large volumes of data and programs not currently being accessed by a processor.

bank conflict: A bank "busy-wait" situation. Since memory chip speeds are relatively slow when required to deliver a single word, supercomputer memories are placed in a large number of independent banks (usually a power of 2). A vector laid out contiguously in memory (one component per successive bank) can be accessed at one word per cycle (despite the intrinsic slowness of memory chips) through the use of pipelined delivery of vector-component words at high bandwidth. When the number of banks is a power of 2, then vectors requiring strides of a power of 2 can run into a bank conflict.

bank cycle time: The time, measured in clock cycles, taken by a memory bank between the honoring of one request to fetch or store a data item and accepting another such request. On most supercomputers this value is either four or eight clock cycles.

BLAS—Basic Linear Algebra Subprograms, a set of Fortran-callable subroutines that perform "kernel" linear algebra operations. Three levels of BLAS currently exist.

barrier synchronization: A means for synchronizing a set of processors in a shared-memory multiprocessor system by halting processors in that set at a specified barrier point until every processor in the set reaches the barrier. At this point the barrier is "removed," and all processors are allowed to resume execution.

cache: Small interface memory with better fetch speed than main memory. The term is more often used when this memory is required to interface with main memory. Cache memory is usually made transparent to the user. A reference to a given area of main memory for one piece of data or instruction is usually closely followed by several additional references to that same area for other data or instruction. Consequently, a cache is automatically filled by a predefined algorithm. The computer system manages the "prefetch" process.

cache coherence: The state that exists when all caches within a multiprocessor have identical values for any shared variable that is simultaneously in two or more caches.

cache hit: A cache access that successfully finds the requested data.

cache line: The unit in which data is fetched from memory to cache.

cache miss: A cache access that fails to find the requested data. The cache must then be filled from main memory at the expense of time.

chaining (linking): The ability to take the results of one vector operation and use them directly as input operands to a second vector instruction, without the need to store to memory or registers the results of the first vector operation. Chaining two vector floating-point operations, for example, could double the asymptotic Mflops rate.

chime: "Chained vector time," approximately equal to the vector length in a DO-loop. The number of chimes required for a loop dominates the time required for execution. A new chime begins in a loop each time a resource (functional unit, vector register, or memory path) must be reused.

chunksize: The number of iterations of a parallel DO-loop grouped together as a single task in order to increase the granularity of the task.

CISC (Complex Instruction Set Computer): A computer with an instruction set that includes complex (multicycle) instructions.

clock cycle: The fundamental period of time in a computer. For example, the clock cycle of a CRAY-2 is 4.1 nsec.

coarse-grain parallelism: Parallel execution in which the amount of computation per task is significantly larger than the overhead and communication expended per task.

combining switch: An element of an interconnection network that can combine certain types of requests into one request and produce a response that mimics serial execution of the requests.

common subexpression: A combination of operations and operands that is repeated, especially in a loop:

```
DO 20 I = 1, N
   A(I) = 2.0 + B(I) * C(I) + X(I) / T(I)
   Y(I) = P(I) / (2.0 + B(I) * C(I))
   D(I) = X(I) / T(I) + U(I)
20 CONTINUE
```

The following are common subexpressions in the preceding loop:

```
2.0 + B(I) * C(I)
X (I)/T(I)
```

A good compiler will not recompute the common subexpressions but will save them in a register for reuse.

compiler directives: Special keywords specified on a comment card, but recognized by a compiler as providing additional information from the user for use in optimization. For example,

CDIR$ IVDEP

specifies to a CRAY compiler that no data dependencies exist among the array references in the loop following the directive.

compress/index: A vector operation used to deal with the nonzeros of a large vector with relatively few nonzeros. The location of the nonzeros is indicated by an index vector (usually a bit vector of the same length in bits as the full vector in words). The compress operation uses the index vector to gather the nonzeros into a dense vector where they are operated on with a vector instruction. See also **gather/scatter**.

concurrent processing: Simultaneous execution of instructions by two or more processors within a computer.

critical section: A section of a program that can be executed by at most one process at a time.

crossbar (interconnection): An interconnection in which each input is connected to each output through a path that contains a single switching node.

cycle (of computer clock): An electronic signal that counts a single unit of time within a computer.

cycle time: The length of a single cycle of a computer function such as a memory cycle or processor cycle. See also **clock cycle**.

data cache: A cache that holds data but does not hold instructions.

data dependency: The situation existing between two statements if one statement can store into a location that is later accessed by the other statement. For example, in

```
S1: C = A + B
S2: Z = C * X + Y
```

S2 is data dependent on S1: S1 must be executed before S2 so that C is stored before being used in S2. A recursive data dependency involves statements in a DO-loop such that a statement in one iteration depends on the results of a statement from a previous iteration. For example, in

```
    DO 30 I = 1, N
       A(I) = B(I) * A(I-1) + C(I)
 30 CONTINUE
```

a recursive data dependency exists in the assignment statement in loop 30: the value of A(I) computed in one iteration is the value A(I-1) needed in the next iteration.

deadlock: A situation where two or more processes are waiting indefinitely for an event that can be caused only by one of the waiting processes.

dependency analysis: An analysis (by compiler or precompiler) that reveals which portions of a program depend on the prior completion of other portions of the program. Dependency analysis usually relates statements in a DO-loop.

direct mapping: A cache that has a set associativity of one, so that each item has a unique place in the cache at which it can be stored.

disk striping: Interleaving a disk file across two or more disk drives to enhance input/output performance. The performance gain is a function of the number of drives and channels used.

distributed memory: A form of storage in which each processor can access only a part of the total memory in the system and must explicitly communicate with other processors in the system to share data.

distributed processing: Processing on a network of computers that do not share main memory.

explicitly parallel: Language semantics that describe which computations are independent and can be performed in parallel. See also **implicitly parallel**.

fetch-and-add: A computer synchronization instruction that updates a word in memory, returns the value before the update, and (if executed concurrently by several processors simultaneously) produces a set of results as if the processors executed in some arbitrary order.

fine-grain parallelism: A type of parallelism where the amount of work per task is relatively small compared with the overhead necessary for communication and synchronization.

flops: Arithmetic floating-point operations (addition and multiplication) per second.

functional units: Functionally independent parts of the ALU each of which performs a specific function, for example, address calculation, floating-point add, and floating-point multiply.

gather/scatter: The operations related to large, sparse data structures. A full vector with relatively few nonzeros is transformed into a vector with only those nonzeros by using a gather operation. The full vector, or one with the same structure, is built from the inverse operation or scatter. The process is accomplished with an index vector, which is usually the length of the number of nonzeros, with each component being the relative location in the full vector. See also **compress/index**.

GAXPY: A generalized *SAXPY* operation, taking linear combinations of a set of columns and accumulating the columns in a vector, as in matrix-vector product.

Gflops (Gigaflops): A computation rate of one billion floating-point operations per second.

global memory: A memory accessible to all of the computer's processors.

granularity: The size of the tasks to be executed in parallel. Fine granularity is illustrated by execution of statements or small loop iterations as separate processes; coarse granularity involves subroutines or sets of subroutines as a separate process. The greater the number of processes, the "finer" the granularity and the greater the overhead required to keep track of them.

Grosch's law: An empirical rule that the cost of computer systems increases as the square root of the computational power of the systems.

hierarchy (of memory system): A multilevel memory structure in which successive levels are progressively larger, slower, and cheaper. Examples are registers, cache, local memory, main memory, secondary storage, and disk.

high-speed buffer memory: A part of the memory that holds data being transferred between a large main memory and the registers of high-speed processors.

hit ratio: the ratio of the number of times data requested is found in the cache.

hot-spot contention: An interference phenomenon observed in multiprocessors caused by memory access statistics being slightly skewed from a uniform distribution to favor a specific memory module.

hypercube architecture: A local-memory architecture where the processors are connected in a topology known as a hypercube. For a hypercube with 2^d processors, each CPU has d nearest neighbors.

implicitly parallel: Language semantics that do not allow the user to explicitly describe which computations are independent and can be performed in parallel. For an implicitly parallel language, the compiler must deduce or prove independence in order to make use of parallel hardware. The comparative difficulty of the deduction separates implicitly parallel languages from explicitly parallel languages.

instruction buffer: A small, high-speed memory that holds instructions recently executed or about to be executed.

instruction cache: A cache memory that holds only instructions, but not data.

instruction scheduling: A strategy of a compiler to analyze the outcome of the operations specified in a program and to issue instructions in an optimal manner. That is, the instructions are not necessarily issued in the order specified by the programmer, but in an order that optimally uses the registers, functional units, and memory paths of the computer—at the same time guaranteeing correct results for the computation.

instruction set: The set of low-level instructions that a computer is capable of executing. Programs expressed in a high-level language must ultimately be reduced to these.

interactive vectorizer: An interactive program to help a user vectorize source code. The program analyzes the source for loops and operation sequences that can be accomplished by using vector instructions or macros. When obstructions to vectorization are found, the user is informed. Often the user can indicate that a vector reference with a potential recursive reference is "safe," or the user can remove an IF-test, branch, or subroutine call, in order to achieve vectorization. See also **true ratio.**

interconnection network: The system of logic and conductors that connects the processors in a parallel computer system. Some examples are bus, mesh, hypercube, and Omega networks.

interprocessor communication: The passing of data and information among the processors of a parallel computer during the execution of a parallel program.

interprocessor contention: Conflicts caused when multiple CPUs compete for shared system resources. For example, memory bank conflicts for a user's code in global-memory architectures are caused by other processors running independent applications.

invariant: A variable, especially in a DO-loop, that appears only on the right side of an equals signs. The variable is read only; it is never assigned a new value.

invariant expression: An expression, especially in a DO-loop, all of whose operands are invariant or constant.

linking: See **chaining.**

load balancing: A strategy in which the longer tasks and shorter tasks are carefully allocated to different processors to minimize synchronization costs or task startup overhead.

local memory: A form of storage in which communication with the small, fast memory is under user control. Local memory is similar to cache, but under explicit program control.

locality: A pragmatic concept having to do with reducing the cost of communication by grouping together objects. Certain ways of laying out data can lead to faster programs.

lock: A shared-data structure that multiple processes access to compete for the exclusive right to continue execution only if no other processor currently holds that exclusive right. Locks are typically implemented as primitive operations.

loop unrolling: A loop optimization technique for both scalar and vector architectures. The iterations of an inner loop are decreased by a factor of two or more by explicit inclusion of the very next one or several iterations. Loop unrolling can allow traditional scalar compilers to make better use of registers and to improve overlap operations. On vector machines, loop unrolling may either improve or degrade performance; the process involves a tradeoff between overlap and register use on the one hand and vector length on the other.

Mflops (megaflops): Millions of floating-point (arithmetic) operations per second, a common rating of supercomputers and vector instruction machines.

main memory: A level of random-access memory, the primary memory of a computer. Typical sizes for large computers are 64–2048 Mbytes.

macrotasking: Dividing a computation into two or more large tasks to be executed in parallel. Typically the tasks are subroutine calls executed in parallel.

memory bank conflict: A condition that occurs when a memory unit receives a request to fetch or store a data item prior to completion of its bank cycle time since its last such request.

memory management: The process of controlling the flow of data among the levels of memory hierarchy.

microtasking: Parallelism at the DO-loop level. Different iterations of a loop are executed in parallel on different processors.

MIMD: A multiple-instruction stream/multiple-data stream architecture. Multiple-instruction streams in MIMD machines are executed simultaneously. MIMD terminology is used more generally to refer to multiprocessor machines. See also **SIMD, SISD.**

mini-supercomputer: A machine with roughly one-tenth to one-half the performance capability of supercomputer and available at roughly one-tenth the price. "Mini-supers" use a blend of minicomputer technology and supercomputer architecture (pipelining, vector instructions, parallel CPUs) to achieve attractive price and performance characteristics.

MIPS: Millions of instructions per second.

MOPS: Millions of operations per second.

multiprocessing: The ability of a computer's operating system to mix separate user jobs on one or more CPUs. See also **multitasking.**

multiprocessor: A computer system with more than one CPU. The CPUs are usually more tightly coupled than simply sharing a local network. For example, a system with CPUs that use a common bus to access a shared memory is called a multiprocessor.

multiprogramming: The ability of a computer to time share its CPU with more than one program on a CPU. See also **multitasking.**

multitasking: The execution of multiple tasks from the same program on one or more processors. The terms multiprogramming, multiprocessing, and multitasking are often used interchangeably (with a notable lack of precision) to describe three different concepts. We use multiprogramming to refer to the ability of a computer to share more than one program on at least one CPU, and multitasking to refer the simultaneous execution of several tasks from the same program on two or more CPUs, or the ability of a computer to intermix jobs on one or more CPUs.

$n_{1/2}$: The vector length required to achieve one-half the peak performance rate. A large value of $n_{1/2}$ indicates severe overhead associated with vector startup. The rule of thumb is that if the average vector length is 3 times $n_{1/2}$, there is great efficiency; if the vector length is less than that, there is inefficiency.

optimization: A process whereby a compiler tries to make the best possible use of the target computer's hardware to perform the operations specified by a programmer. Alternatively, optimization refers to the process whereby a programmer tries to make optimal use of the target programming language to produce optimal code for the computer architecture.

optimization block: A block of code (rarely a whole subprogram, often a single DO-loop) in which a compiler optimizes the generated code. A few compilers attempt to optimize across such blocks; many work on each block independently.

page: The smallest managed unit of a virtual memory system. The system maintains separate virtual-to-physical translation information for each page.

parallel computer: A computer that can perform multiple operations simultaneously, usually because multiple processors (that is, control units or ALUs) or memory units are available. Parallel computers containing a small number (say, less than 50) of processors are called multiprocessors; those with more than 1000 are often referred to as massively parallel computers.

parallel processing: Processing with more than one CPU on a single program simultaneously.

parallelization: The simultaneous execution of separate parts of a single program.

parsing: The process whereby a compiler analyzes the syntax of a program to establish the relationships among operators, operands, and other tokens of a program. Parsing does not involve any semantic analysis.

partitioning: Restructuring a program or algorithm into independent computational segments. The goal is to have multiple CPUs simultaneously work on the independent computational segments.

percentage parallelization: The percentage of CPU expenditure processed in parallel on a single job. It is usually not possible to achieve 100 percent of an application's processing time to be shared equally on all CPUs.

percentage vectorization: The percentage of an application executing in vector mode. This percentage may be calculated as a percentage of CPU time or as the percentage of lines of code (usually Fortran) in vector instructions. The two approaches are not consistent and may give very different percentage ratings. The first calculation method leads to performance improvement as measured by CPU time, while the second method measures the "success rate" of the compiler in converting scalar code to vector code. The former is, of course, the more meaningful hardware performance measure. See also **vectorization**.

perfect-shuffle interconnection: An interconnection structure that connects processors according to a permutation that corresponds to a perfect shuffle of a deck of cards.

physical memory: The actual memory of a computer directly available for fetching or storing of data (contrast with virtual memory).

pipelining: The execution of a set of operations, by a single processor, such that subsequent operations in the sequence can begin execution before previous ones have completed execution. Pipelining is an assembly line approach to data or instruction execution. In modern supercomputers the floating-point operations are often pipelined with memory fetches and stores of the vector data sets.

primary memory: Main memory accessible by the CPU(s) without using input/output processes. See also **secondary memory**.

process: A sequential program in execution. A program by itself is not a process; a program is a passive entity, while a process is an active entity.

pseudovector: A scalar temporary.

r_∞: The asymptotic rate for a vector operation as the vector length approaches infinity.

recurrence: A relationship in a DO-loop whereby a computation in one iteration of the loop depends upon completion of a previous iteration of the loop. Such dependencies inhibit vectorization.

reduced instruction set computer (RISC): A philosophy of instruction set design where a small number of simple, fast instructions are implemented rather than a larger number of slower, more complex instructions.

reduction function: A special type of recurrence relationship where a vector of values is reduced to a single scalar result. The variable containing the result is referred to as a "reduction-function scalar." Most compilers recognize several common reduction functions, such as sum, dot product, the minimum (maximum) of the elements of a vectorizable expression, and variants on these themes.

scalar processing: The execution of a program by using instructions that can operate only on a single pair of operands at a time (contrast with vector processing).

scalar temporary: A compiler-created scalar variable that holds the value of a vectorizable expression on each iteration of a DO-loop.

scoreboard: A hardware device that maintains the state of machine resources to enable instructions to execute without conflict at the earliest opportunity.

secondary memory: A larger and slower memory than primary memory. Access to it often requires special instructions, such as I/O instructions. See also **primary memory**.

semaphore: A variable that is used to control access to shared data.

set associative: A cache structure in which all tags in a particular set are compared with an access key in order to access an item in cache. The set may have as few as one element or as many elements as there are lines in the full cache.

SIMD: single instruction stream/multiple data stream architecture. Currently, three such machines dominate the market: the AMT DAP, Goodyear, and the Connection Machine.

SISD: Single instruction stream/single data stream system. Instructions are processed sequentially, with the flow of data from memory to the functional unit and back to memory—the traditional configuration of computers.

speedup: The factor of performance improvement over pure scalar performance. The term is usually applied to performance of either one CPU versus multiple CPUs or vector versus scalar processing. The reported numbers are often misleading because of an inconsistency in reporting the speedup over an original or revised process running in scalar mode.

spin lock: An implementation of the LOCK primitive that causes a processor to retest a semaphore until it changes value. Busy waits will spin until the lock is free.

strength reduction: A process whereby a compiler attempts to replace instructions with less costly instructions that produce identical results. For example, X**2 becomes X*X.

stride: A term derived from the concept of walking (striding) through the data from one location to the next. The term is often used in relation to vector storage. A mathematical vector is represented in a computer as an array of numbers. Most computers use contiguous storage of vectors, locating them by the address of the first word and vector length. Many applications in linear algebra, however, require load and store of vectors with components that do not reside contiguously in memory, such as the rows of a matrix stored in column order. The row elements are spaced in memory by a distance (stride) of the leading dimension of the array representing the matrix. Some vector computers allow vector fetching and storing to occur with randomly stored vectors. An index vector locates successive components. This is often useful in storing the nonzero elements of a sparse vector. See also **vector**.

stripmining: A process used by a compiler on a register-to-register vector processor whereby a DO-loop of long or variable iteration count is executed in "strips" which are the length of a vector register, except for a "remainder" strip whose length is less. For example, on a Cray computer with a vector length of 64, a loop of iteration count 150 is performed in one strip of length 22 (the remainder) and two strips of length 64.

supercomputer(s): At any given time, that class of general-purpose computers that are faster than their commercial competitors and have sufficient central memory to store the problem sets for which they are designed. Computer memory, throughput, computational rates, and other related computer capabilities contribute to performance. Consequently, a quantitative measure of computer power in large-scale scientific processing does not exist, and a precise definition of supercomputers is difficult to formulate.

superword: A term used on the CYBER 205 and ETA 10 to describe a group of eight 64-bit words, or, alternatively, sixteen 32-bit "half-words." The memory units on these machines generally fetch and store data in superwords (also called "swords"), regardless of the size of the data item referenced by the user program.

synchronization: An operation in which two or more processors exchange information to coordinate their activity.

test and set: An instruction that typically tests a memory location (flag/semaphore) and updates it according to the flag's value. It is atomic in that after the flag is read and before the flag is updated, the CPU executing the instruction will not allow access to the flag.

thrashing: A phenomenon of virtual memory systems that occurs when the program, by the manner in which it is referencing its data and instructions, regularly causes the next memory locations referenced to be overwritten by recent or current instructions. The result is that referenced items are rarely in the machine's physical memory and almost always must be fetched from secondary storage, usually a disk.

thread: A lightweight or small-granularity process.

translation look-aside buffer (TLB): The TLB is the memory cache of the most recently used page table entries within the memory management unit.

true ratio: The frequency with which the "true" branch in a Fortran IF-test occurs. If the true ratio is known at compile time, some compilers can take advantage of this knowledge. However, the true ratio is often data dependent and cannot effectively be dealt with automatically. See also **interactive vectorizer.**

ultracomputer: A shared-memory MIMD computer incorporating a perfect-shuffle interconnection network capable of combining colliding message packets according to simple logical and mathematical functions.

unneeded store: The situation resulting when two or more stores into the same memory location without intermediate reads occur in an optimization block, especially within a DO-loop, such that only the last store need actually be performed. For example, in the loop below, the calculation of A(I) in the first statement will be overwritten by the calculation for A(I) in the last statement. Consequently, A(I) need not be stored when calculated in the first statement but must be kept in a register for the second and third statement.

```
    DO 40 I + 1, N
       A(I) = B(I) * C(I) / (E(I) + F(I))
       X(I) = Y(I) * A(I)
       Z(I) = R(I) + Q(I) * A(I)
       A(I) = X(I) + Y(I) * Z(I)
 40 CONTINUE
```

vector: An ordered list of items in a computer's memory. A simple vector is defined as having a starting address, a length, and a stride. An indirect address vector is defined as having a relative base address and a vector of values to be applied as indexes to the base. Consider the following:

```
    DO 50 I = 1, N
       J = J * J / I
       K = K + 2
       A(I) = B(IB(I)) * C(K) + D(J)
 50 CONTINUE
```

All of these vectors have length N; A and C are vectors with strides of one and two, respectively; B is an indirect address vector with the simple vector IB holding the indexes; and the vector of indirect address indexes of D can be computed at execution time from the initial value of J.

vector processing: A mode of computer processing that acts on operands that are vectors or arrays. Modern supercomputers achieve speed through pipelined arithmetic units. Pipelining, when coupled with instructions designed to process all the elements of a vector rather than one data pair at a time, is known as vector processing.

vector register: A storage device that acts as an intermediate memory between a computer's functional units and main memory.

vectorization: The act of tuning an application code to take advantage of vector architecture. See also **percentage vectorization.**

vectorize: The process whereby a compiler or programmer generates vector instructions for a loop. Vector instructions are typically in the form of machine-specific instructions but can also be expressed as calls to special subroutines or in an array notation.

virtual memory: A memory scheme that provides a programmer with a larger memory than that physically available on a computer. As data items are referenced by a program, the system assigns them to actual physical memory locations. Infrequently referenced items are transparently migrated to and from secondary storage—often, disks. The collection of physical memory locations assigned to a program is its "working set."

virtual processor: An abstraction away from the physical processors of a computer that permits one to program the computer as if it had more processors than are actually available. The physical processors are time-shared among the virtual processors.

VLIW (very long instruction word): Reduced-instruction-set machines with a large number of parallel, pipelined functional units but only a single thread of control. Instead of coarse-grain parallelism of vector machines and multiprocessors, VLIWs provide fine-grain parallelism.

VLSI (very large-scale integration): A manufacturing process that uses a fixed number of manufacturing steps to produce all components and interconnections for hundreds of devices each with millions of transistors.

von Neumann bottleneck: The data path between the processor and memory of a computer that most constrains performance of such a computer.

von Neumann computer: An SISD computer in which one instruction at a time is decoded and performed to completion before the next instruction is decoded. A key point of a von Neumann computer is that both data and program live in the same memory.

working set: See **virtual memory.**

wrap-around scalar: A scalar variable whose value is set in one iteration of a DO-loop and referenced in a subsequent iteration and is consequently recursive. All common reduction-function scalars are wrap-around scalars and usually do not prevent vectorization. All other wrap-around scalars usually do prevent vectorization of the loop in which they appear. In the following, all scalars wrap around except S.

```
        DO 60 I = 1, N
          S = T
          T = A(I) * B(I)
          SUM = SUM + T/S
          IF (T.GT.0) THEN
              Q = X(I) + Y(I) / Z(I)
          ENDIF
          R(I) = Q + P(I)
     60 CONTINUE
```

The scalar Q is a wrap around because on any iteration for which T is less than or equal to 0 is not true, the value used to compute $R(I)$ wraps around from the previous iteration.

write-in cache: A cache in which writes to memory are stored in the cache and written to memory only when a rewritten item is removed from cache. This is also referred to as write-back cache.

write-through cache: A cache in which writes to memory are performed concurrently both in cache and in main memory.

Appendix C

Information on Various High-Performance Computers

C1. The Major Supercomputers

Cray Research and Cray Computer Corp.

Machine	Cycle Time	No. of Proc.	Memory (Mwords)	Peak Perf.	Year Shipped
CRAY-1 †	12.5	1	1	160	1976
CRAY-1/S †	12.5	1	4	160	1979
CRAY X-MP †	9.5	2	4	420	1982
CRAY X-MP †	9.5	4	8	840	1984
CRAY X-MP †	9.5	2	16	840	1985
CRAY-2	4.1	4	256	194	1985
CRAY X-MP †	8.5	4	16	940	1986
CRAY X-MP †	8.5	4	32	940	1988
CRAY Y-MP	6.0	8	32	2700	1988
CRAY-3 ‡	2	16	512+	16000	1991
CRAY C-90 ‡	4	16	64+	16000	1991
CRAY-4 ‡	1	64	1000+	128000	199?

†Machines no longer manufactured.

‡Machines not yet available.

ETA Systems

Machine	Cycle Time	Memory (Mwords)	Peak Perf.	Year Released
ETA-10P †	24	16	375	1987
ETA-10E †	10.5	128	1700	1987
ETA-10Q †	19	16	475	1988
ETA-10G †	7	256	5150	1988

†Machines no longer manufactured.

Fujitsu

Machine	Cycle Time	Memory (Mwords)	Peak Perf.	Year Shipped
VP-2600/10	3.2		5000	1990
VP-100E	7	64	429	1984
VP-200E	7	128	857	1984
VP-50E	7	64	286	1985
VP-400E	7	128	1700	1986
VP-30E	7.5	32	133	1987

Hitachi

Machine	Cycle Time	Memory (Mwords)	Peak Perf.	Year Shipped
S-810/10	14	16	315	1983
S-810/20	14	32	620	1983
S-810/5	14	16	160	1986
S-820/60	4	32	1500	1988
S-820/80	4	64	3000	1988

IBM

Machine	Cycle Time	Memory (Mwords)	Peak Perf.	Year Shipped
3090/180 VF	18.5	8	108	1986
3090/150E VF	17.75	8	112	1987
3090/600E VF	17.2	32	696	1988
3090/500E VF	17.2	32	580	1988
3090/400E VF	17.2	32	464	1988
3090/300E VF	17.2	16	348	1988
3090/200E VF	17.2	16	232	1988
3090/280E VF	17.2	16	232	1988
3090/180E VF	17.2	8	116	1988
3090/120E VF	18.5	8	108	1988
3090/600S VF	15	32	800	1988
3090/500S VF	15	32	666	1988
3090/400S VF	15	32	533	1988
3090/300S VF	15	16	400	1988
3090/200S VF	15	16	266	1988
3090/180S VF	15	8	133	1988
3090/600J VF	14.5	64	828	1989
3090/500J VF	14.5	64	690	1989
3090/400J VF	14.5	64	552	1989
3090/300J VF	14.5	32	414	1989
3090/200J VF	14.5	32	275	1989
3090/180J VF	14.5	16	138	1989

NEC

Machine	Cycle Time	Memory (Mwords)	Peak Perf.	Year Shipped
SX2-100	6	16	285	1980
SX2-400	6	32	1300	1985
SX2-200	6	32	570	1986
SX3-44	2.9	200	22000	1990-91

C2. The Major Mini-supercomputers

Machine System	Peak Mflops	Year Intro.
Convex C1	20	1984
Alliant FX-Series	94	1985
Alliant FX-80	188	1987
Convex C2	200	1987
DEC 9000	500	1990

C3. Machine Categories

Companies Building High-Performance Computers

Available	Supers	Vector M-frames	Mini-supers
	CRAY	CDC	Alliant
	Fujitsu	Fujitsu	Convex
	Hitachi	IBM	DEC
	NEC	NAS	
		Unisys	
		Hitachi	
		Honeywell	
Being Built	SSI (Chen)		
	KSR		
	Tera		
No Longer in Business	Chopp	Amer. Super	Culler
	Denelcor	Cydrome	SAXPY
	ETA	Prisma	Astronautics
	Trilogy	SCS	Vitesse
		Key	Gould
			Multiflow
			Supertek

Available	mP	mC	SIMD	Graphics
	Arete	NCUBE	AMT	Apollo
	BBN	Cogent (t)	MasPar	SGI
	Concurrent	Intel iPSC	TMC	Stardent
	Encore	Meiko (t)		
	IP1	Paralax		
	Masscomp	Parsytec (t)		
	Myrias	Suprenum		
	Plexus			
	Sequent			
Being Built				
No Longer in Business	E&S	Symult		
	Elxsi	Topologics (t)		
	Flexible	FPS T-Series		
	Synapse			

C5. Comparison of Supercomputer Technology

Computer	Cooling Technology
CRAY Y-MP	Liquid
CRAY X-MP	Freon
CRAY-2	Liquid Immersion
ETA-10E	Liquid Nitrogen Immersion/Logic Chilled Air/Memory
Fujitsu VP-200	Air Cooled
Hitachi S-810/20	Air/Memory & Water/Logic
IBM 3090/VF	Liquid; water
NEC SX/2	Water

Main Memory Sizes

Computer System	No. of. Proc.	Size, MW	No. of Banks	Bank Cycles	Wait Time
CRAY-1	1	4	16	4	50
CRAY X-MP *ECL*	1-4	16	64	4	34
CRAY X-MP *MOS*	1-4	64	64	8	68
CRAY Y-MP	1-8	32	256	5	30
CRAY-2	1-4	512	128	46	184
CRAY-2S	1-4	128	128	13	53
CYBER 205	1	64	8		
ETA-10E	1-8	128	64	20	214
Fujitsu VP-200	1	32	128	8	55
Fujitsu VP-400	1	32	256	8	55
Hitachi S-820	1	32	128	17	70
IBM 3090/VF	1-6	16	16		
NEC SX/2	1	32	512	7	40

Registers and Buffer Size

Computer System	Single CPU Register Configuration (64-bit words)
CRAY-1	8 × 64
CRAY X-MP	8 × 64
CRAY Y-MP	8 × 64
CRAY-2	8 × 64 plus 16K local memory/processor
CYBER 205	buffer
ETA-10	buffer
Fujitsu VP-200	reconfigurable 8 × 1024 . . . 256 × 16
Fujitsu VP-400	reconfigurable 16 × 1024 . . . 512 × 16
Hitachi S-820	32 × 256
IBM 3090/VF	8 × 128 plus 8K cache/processor
NEC SX-2	40 × 256

Paths to Memory

Computer System	No. of Paths to Memory	No. of Paths/ No. of Fl. Pt.	Latency, cycles
CRAY-1	1	.5	15
CRAY X-MP	3	1.5	14
CRAY Y-MP	3	1.5	17
CRAY-2	1	.5	35-50
CYBER 205	3	1.5	50
ETA-10	3	1.5	
Fujitsu VP-200	1*	.5	31-33
Fujitsu VP-400	1	.5	31-33
IBM 3090/VF	1	.5	
NEC SX/2	12	.75	

*The Fujitsu VP-200 has 2 paths for contiguously stored vectors.

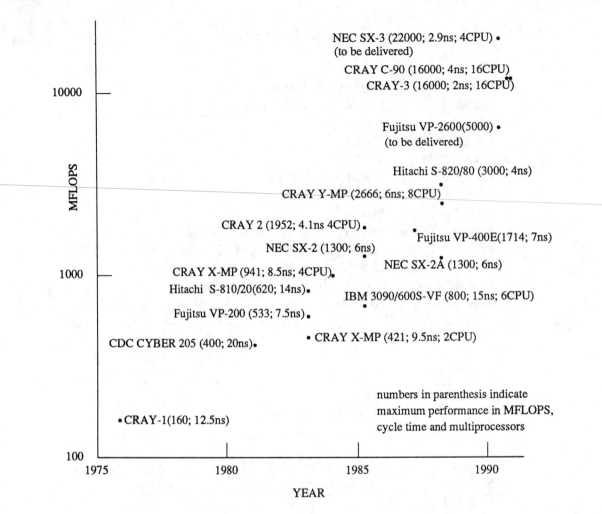

Figure C.1: **Supercomputers over Time**

Appendix D

Level 1, 2, and 3 BLAS Quick Reference

Level 1 BLAS

```
                  dim scalar vector   vector   scalars        5-element  prefixes
                                                              array
SUBROUTINE _ROTG (                                A, B, C, S )         S, D
SUBROUTINE _ROTMG(                        D1, D2, A, B,        PARAM )  S, D
SUBROUTINE _ROT  ( N,         X, INCX, Y, INCY,          C, S )        S, D
SUBROUTINE _ROTM ( N,         X, INCX, Y, INCY,               PARAM )  S, D
SUBROUTINE _SWAP ( N,         X, INCX, Y, INCY )                       S, D, C, Z
SUBROUTINE _SCAL ( N,  ALPHA, X, INCX )                                S, D, C, Z, CS, ZD
SUBROUTINE _COPY ( N,         X, INCX, Y, INCY )                       S, D, C, Z
SUBROUTINE _AXPY ( N,  ALPHA, X, INCX, Y, INCY )                       S, D, C, Z
FUNCTION    _DOT ( N,         X, INCX, Y, INCY )                       S, D, DS
FUNCTION    _DOTU ( N,        X, INCX, Y, INCY )                       C, Z
FUNCTION    _DOTC ( N,        X, INCX, Y, INCY )                       C, Z
FUNCTION    __DOT ( N, ALPHA, X, INCX, Y, INCY )                       SDS
FUNCTION    _NRM2 ( N,        X, INCX )                                S, D, SC, DZ
FUNCTION    _ASUM ( N,        X, INCX )                                S, D, SC, DZ
FUNCTION    I_AMAX( N,        X, INCX )                                S, D, C, Z
```

Level 2 BLAS

```
        options              dim  b-width scalar matrix  vector   scalar vector  prefixes

_GEMV (         TRANS,    M, N,         ALPHA, A, LDA, X, INCX, BETA, Y, INCY ) S, D, C, Z
_GBMV (         TRANS,    M, N, KL, KU, ALPHA, A, LDA, X, INCX, BETA, Y, INCY ) S, D, C, Z
_HEMV ( UPLO,             N,            ALPHA, A, LDA, X, INCX, BETA, Y, INCY ) C, Z
_HBMV ( UPLO,             N, K,         ALPHA, A, LDA, X, INCX, BETA, Y, INCY ) C, Z
_HPMV ( UPLO,             N,            ALPHA, AP,    X, INCX, BETA, Y, INCY ) C, Z
_SYMV ( UPLO,             N,            ALPHA, A, LDA, X, INCX, BETA, Y, INCY ) S, D
_SBMV ( UPLO,             N, K,         ALPHA, A, LDA, X, INCX, BETA, Y, INCY ) S, D
_SPMV ( UPLO,             N,            ALPHA, AP,    X, INCX, BETA, Y, INCY ) S, D
_TRMV ( UPLO, TRANS, DIAG,   N,                A, LDA, X, INCX )               S, D, C, Z
_TBMV ( UPLO, TRANS, DIAG,   N, K,             A, LDA, X, INCX )               S, D, C, Z
_TPMV ( UPLO, TRANS, DIAG,   N,                AP,    X, INCX )               S, D, C, Z
_TRSV ( UPLO, TRANS, DIAG,   N,                A, LDA, X, INCX )               S, D, C, Z
_TBSV ( UPLO, TRANS, DIAG,   N, K,             A, LDA, X, INCX )               S, D, C, Z
_TPSV ( UPLO, TRANS, DIAG,   N,                AP,    X, INCX )               S, D, C, Z


        options              dim  scalar vector   vector   matrix  prefixes

_GER  (              M, N, ALPHA, X, INCX, Y, INCY, A, LDA ) S, D
_GERU (              M, N, ALPHA, X, INCX, Y, INCY, A, LDA ) C, Z
_GERC (              M, N, ALPHA, X, INCX, Y, INCY, A, LDA ) C, Z
_HER  ( UPLO,           N, ALPHA, X, INCX,          A, LDA ) C, Z
_HPR  ( UPLO,           N, ALPHA, X, INCX,          AP )     C, Z
_HER2 ( UPLO,           N, ALPHA, X, INCX, Y, INCY, A, LDA ) C, Z
_HPR2 ( UPLO,           N, ALPHA, X, INCX, Y, INCY, AP )     C, Z
_SYR  ( UPLO,           N, ALPHA, X, INCX,          A, LDA ) S, D
_SPR  ( UPLO,           N, ALPHA, X, INCX,          AP )     S, D
_SYR2 ( UPLO,           N, ALPHA, X, INCX, Y, INCY, A, LDA ) S, D
_SPR2 ( UPLO,           N, ALPHA, X, INCX, Y, INCY, AP )     S, D
```

Level 3 BLAS

```
        options                    dim        scalar matrix  matrix  scalar matrix  prefixes

_GEMM (         TRANSA, TRANSB,  M, N, K, ALPHA, A, LDA, B, LDB, BETA, C, LDC ) S, D, C, Z
_SYMM ( SIDE, UPLO,              M, N,    ALPHA, A, LDA, B, LDB, BETA, C, LDC ) S, D, C, Z
_HEMM ( SIDE, UPLO,              M, N,    ALPHA, A, LDA, B, LDB, BETA, C, LDC ) C, Z
_SYRK (       UPLO, TRANS,          N, K, ALPHA, A, LDA,         BETA, C, LDC ) S, D, C, Z
_HERK (       UPLO, TRANS,          N, K, ALPHA, A, LDA,         BETA, C, LDC ) C, Z
_SYR2K(       UPLO, TRANS,          N, K, ALPHA, A, LDA, B, LDB, BETA, C, LDC ) S, D, C, Z
_HER2K(       UPLO, TRANS,          N, K, ALPHA, A, LDA, B, LDB, BETA, C, LDC ) C, Z
_TRMM ( SIDE, UPLO, TRANSA,   DIAG, M, N,    ALPHA, A, LDA, B, LDB )            S, D, C, Z
_TRSM ( SIDE, UPLO, TRANSA,   DIAG, M, N,    ALPHA, A, LDA, B, LDB )            S, D, C, Z
```

Name	Operation	Prefixes
_ROTG	Generate plane rotation	S, D
_ROTMG	Generate modified plane rotation	S, D
_ROT	Apply plane rotation	S, D
_ROTM	Apply modified plane rotation	S, D
_SWAP	$x \leftrightarrow y$	S, D, C, Z
_SCAL	$x \leftarrow \alpha x$	S, D, C, Z, CS, ZD
_COPY	$y \leftarrow x$	S, D, C, Z
_AXPY	$y \leftarrow \alpha x + y$	S, D, C, Z
_DOT	$dot \leftarrow x^T y$	S, D, DS
_DOTU	$dot \leftarrow x^T y$	C, Z
_DOTC	$dot \leftarrow x^H y$	C, Z
__DOT	$dot \leftarrow \alpha + x^T y$	SDS
_NRM2	$nrm2 \leftarrow \|x\|_2$	S, D, SC, DZ
_ASUM	$asum \leftarrow \|re(x)\|_1 + \|im(x)\|_1$	S, D, SC, DZ
I_AMAX	$amax \leftarrow 1^{st} k \ni |re(x_k)| + |im(x_k)|$ $\quad = max(|re(x_i)| + |im(x_i)|)$	S, D, C, Z

Name	Operation	Prefixes
_GEMV	$y \leftarrow \alpha A x + \beta y, y \leftarrow \alpha A^T x + \beta y, y \leftarrow \alpha A^H x + \beta y, A - m \times n$	S, D, C, Z
_GBMV	$y \leftarrow \alpha A x + \beta y, y \leftarrow \alpha A^T x + \beta y, y \leftarrow \alpha A^H x + \beta y, A - m \times n$	S, D, C, Z
_HEMV	$y \leftarrow \alpha A x + \beta y$	C, Z
_HBMV	$y \leftarrow \alpha A x + \beta y$	C, Z
_HPMV	$y \leftarrow \alpha A x + \beta y$	C, Z
_SYMV	$y \leftarrow \alpha A x + \beta y$	S, D
_SBMV	$y \leftarrow \alpha A x + \beta y$	S, D
_SPMV	$y \leftarrow \alpha A x + \beta y$	S, D
_TRMV	$x \leftarrow A x, x \leftarrow A^T x, x \leftarrow A^H x$	S, D, C, Z
_TBMV	$x \leftarrow A x, x \leftarrow A^T x, x \leftarrow A^H x$	S, D, C, Z
_TPMV	$x \leftarrow A x, x \leftarrow A^T x, x \leftarrow A^H x$	S, D, C, Z
_TRSV	$x \leftarrow A^{-1} x, x \leftarrow A^{-T} x, x \leftarrow A^{-H} x$	S, D, C, Z
_TBSV	$x \leftarrow A^{-1} x, x \leftarrow A^{-T} x, x \leftarrow A^{-H} x$	S, D, C, Z
_TPSV	$x \leftarrow A^{-1} x, x \leftarrow A^{-T} x, x \leftarrow A^{-H} x$	S, D, C, Z

Name	Operation	Prefixes
_GER	$A \leftarrow \alpha x y^T + A, A - m \times n$	S, D
_GERU	$A \leftarrow \alpha x y^T + A, A - m \times n$	C, Z
_GERC	$A \leftarrow \alpha x y^H + A, A - m \times n$	C, Z
_HER	$A \leftarrow \alpha x x^H + A$	C, Z
_HPR	$A \leftarrow \alpha x x^H + A$	C, Z
_HER2	$A \leftarrow \alpha x y^H + y(\alpha x)^H + A$	C, Z
_HPR2	$A \leftarrow \alpha x y^H + y(\alpha x)^H + A$	C, Z
_SYR	$A \leftarrow \alpha x x^T + A$	S, D
_SPR	$A \leftarrow \alpha x x^T + A$	S, D
_SYR2	$A \leftarrow \alpha x y^T + \alpha y x^T + A$	S, D
_SPR2	$A \leftarrow \alpha x y^T + \alpha y x^T + A$	S, D

Name	Operation	Prefixes
_GEMM	$C \leftarrow \alpha op(A)op(B) + \beta C, op(X) = X, X^T, X^H, C - m \times n$	S, D, C, Z
_SYMM	$C \leftarrow \alpha AB + \beta C, C \leftarrow \alpha BA + \beta C, C - m \times n, A = A^H$	S, D, C, Z
_HEMM	$C \leftarrow \alpha AB + \beta C, C \leftarrow \alpha BA + \beta C, C - m \times n, A = A^T$	C, Z
_SYRK	$C \leftarrow \alpha AA^H + \beta C, C \leftarrow \alpha A^H A + \beta C, C - n \times n$	S, D, C, Z
_HERK	$C \leftarrow \alpha AA^T + \beta C, C \leftarrow \alpha A^T A + \beta C, C - n \times n$	C, Z
_SYR2K	$C \leftarrow \alpha AB^H + \alpha BA^H + \beta C, C \leftarrow \alpha A^H B + \alpha B^H A + \beta C, C - n \times n$	S, D, C, Z
_HER2K	$C \leftarrow \alpha AB^T + \alpha BA^T + \beta C, C \leftarrow \alpha A^T B + \alpha B^T A + \beta C, C - n \times n$	C, Z
_TRMM	$B \leftarrow \alpha op(A)B, B \leftarrow \alpha Bop(A), op(A) = A, A^T, A^H, B - m \times n$	S, D, C, Z
_TRSM	$B \leftarrow \alpha op(A^{-1})B, B \leftarrow \alpha Bop(A^{-1}), op(A) = A, A^T, A^H, B - m \times n$	S, D, C, Z

Notes

Meaning of prefixes

S - REAL   C - COMPLEX
D - DOUBLE PRECISION   Z - COMPLEX*16   (this may not be supported
                                                              by all machines)

For the Level 2 BLAS a set of extended-precision routines with the prefixes ES, ED, EC, EZ may also be available.

Level 1 BLAS

In addition to the listed routines there are two further extended-precision dot product routines DQDOTI and DQDOTA.

Level 2 and Level 3 BLAS

Matrix types

GE - GEneral   GB - General Band
SY - SYmmetric   SB - Symmetric Band   SP - Symmetric Packed
HE - HErmitian   HB - Hermitian Band   HP - Hermitian Packed
TR - TRiangular   TB - Triangular Band   TP - Triangular Packed

Options

The dummy options arguments are declared as CHARACTER*1 and may be passed as character strings.

TRANS_ = 'No transpose', 'Transpose', 'Conjugate transpose' (X, X^T, X^H)
UPLO = 'Upper triangular', 'Lower triangular'
DIAG = 'Non-unit triangular', 'Unit triangular'
SIDE = 'Left', 'Right' (A or op(A) on the left, or A or op(A) on the right)

For real matrices, TRANS_ = 'T' and TRANS_ = 'C' have the same meaning.
For Hermitian matrices, TRANS_ = 'T' is not allowed.
For complex symmetric matrices, TRANS_ = 'H' is not allowed.

Appendix E

Operation Counts for Various BLAS and Decompositions

Operation Counts

In this appendix we present the formulas for the operation counts for various BLAS and decompositions.

In the tables below, we give the operation counts for the real dense and banded routines (the counts for the symmetric packed routines are the same as for the dense routines). Separate counts are given for multiplications (including divisions) and additions, and the total is the sum of these expressions. For the complex analogues of these routines, each multiplication would count as 6 operations and each addition as 2 operations, so the total would be different. This information was taken from [7].

Operation Counts for the Level 2 BLAS

The four parameters used in counting operations for the Level 2 BLAS are the matrix dimensions m and n and the upper and lower bandwidths k_u and k_l for the band routines (k if symmetric or triangular). An exact count also depends slightly on the values of the scaling factors α and β, since some common special cases (such as $\alpha = 1$ and $\beta = 0$) can be treated separately.

The count for SGBMV from the Level 2 BLAS is as follows: SGBMV

multiplications:	$mn - (m - k_l - 1)(m - k_l)/2 - (n - k_u - 1)(n - k_u)/2$
additions:	$mn - (m - k_l - 1)(m - k_l)/2 - (n - k_u - 1)(n - k_u)/2$
total flops:	$2mn - (m - k_l - 1)(m - k_l) - (n - k_u - 1)(n - k_u)$

plus m multiplications if $\alpha \neq \pm 1$ and another m multiplies if $\beta \neq \pm 1$ or 0. The other Level 2 BLAS operation counts are shown in Table E.1.

Table E.1: Operation Counts for the Level 2 BLAS

Level 2 BLAS	Multiplications	Additions	Total Flops
SGEMV [1,2]	mn	mn	$2mn$
SSYMV [3,4]	n^2	n^2	$2n^2$
SSBMV [3,4]	$n(2k+1) - k(k+1)$	$n(2k+1) - k(k+1)$	$n(4k+2) - 2k(k+1)$
STRMV [3,4,5]	$n(n+1)/2$	$(n-1)n/2$	n^2
STBMV [3,4,5]	$n(k+1) - k(k+1)/2$	$nk - k(k+1)/2$	$n(2k+1) - k(k+1)$
STRSV [5]	$n(n+1)/2$	$(n-1)n/2$	n^2
STBSV [5]	$n(k+1) - k(k+1)/2$	$nk - k(k+1)/2$	$n(2k+1) - k(k+1)$
SGER [1]	mn	mn	$2mn$
SSYR [3]	$n(n+1)/2$	$n(n+1)/2$	$n(n+1)$
SSYR2 [3]	$n(n+1)$	n^2	$2n^2 + n$

1 – Plus m multiplications if $\alpha \neq \pm 1$
2 – Plus m multiplications if $\beta \neq \pm 1$ or 0
3 – Plus n multiplications if $\alpha \neq \pm 1$
4 – Plus n multiplications if $\beta \neq \pm 1$ or 0
5 – Less n multiplications if matrix is unit triangular

Operation Counts for the Level 3 BLAS

Three parameters are used to count operations for the Level 3 BLAS: the matrix dimensions m, n, and k. In some cases we also must know whether the matrix is multiplied on the left or right. An exact count depends slightly on the values of the scaling factors α and β, but in Table E.2 we assume these parameters are always ± 1 or 0.

Table E.2: **Operation Counts for the Level 3 BLAS**

Level 3 BLAS	Multiplications	Additions	Total Flops
SGEMM	mkn	mkn	$2mkn$
SSYMM (SIDE = 'L')	m^2n	m^2n	$2m^2n$
SSYMM (SIDE = 'R')	mn^2	mn^2	$2mn^2$
SSYRK	$kn(n+1)/2$	$kn(n+1)/2$	$kn(n+1)$
SSYR2K	kn^2	kn^2+n	$2kn^2+n$
STRMM (SIDE = 'L')	$nm(m+1)/2$	$nm(m-1)/2$	nm^2
STRMM (SIDE = 'R')	$mn(n+1)/2$	$mn(n-1)/2$	mn^2
STRSM (SIDE = 'L')	$nm(m+1)/2$	$nm(m-1)/2$	nm^2
STRSM (SIDE = 'R')	$mn(n+1)/2$	$mn(n-1)/2$	mn^2

Operation Counts for Various Decompositions

The parameters used in counting operations are the matrix dimensions m and n, the upper and lower bandwidths k_u and k_l for the band routines (k if symmetric or triangular), and the number of right-hand sides (NRHS) in the solution phase.

LU factorization

multiplications:	$1/2mn^2 - 1/6n^3 + 1/2mn - 1/2n^2 + 2/3n$
additions:	$1/2mn^2 - 1/6n^3 - 1/2mn + 1/6n$
total flops:	$mn^2 - 1/3n^3 - 1/2n^2 + 5/6n$

Computed inverse after the LU factorization

multiplications:	$2/3n^3 + 1/2n^2 + 5/6n$
additions:	$2/3n^3 - 3/2n^2 + 5/6n$
total flops:	$4/3n^3 - n^2 + 5/3n$

Solving systems after LU factorization

multiplications:	NRHS $[n^2]$
additions:	NRHS $[n^2 - n]$
total flops:	NRHS $[2n^2 - n]$

Cholesky factorization

multiplications:	$1/6n^3 + 1/2n^2 + 1/3n$
additions:	$1/6n^3 - 1/6n$
total flops:	$1/3n^3 + 1/2n^2 + 1/6n$

Computed inverse after the Cholesky factorization

multiplications:	$1/3n^3 + n^2 + 2/3n$
additions:	$1/3n^3 - 1/2n^2 + 1/6n$
total flops:	$2/3n^3 + 1/2n^2 + 5/6n$

Solving systems after the Cholesky factorization

multiplications:	$\text{NRHS}\,[n^2 + n]$
additions:	$\text{NRHS}\,[n^2 - n]$
total flops:	$\text{NRHS}\,[2n^2]$

Symmetric indefinite factorization

multiplications:	$1/6n^3 + 1/2n^2 + 10/3n$
additions:	$1/6n^3 - 1/6n$
total flops:	$1/3n^3 + 1/2n^2 + 19/6n$

Computed inverse after the symmetric indefinite factorization

multiplications:	$1/3n^3 + 2/3n$
additions:	$1/3n^3 - 1/3n$
total flops:	$2/3n^3 + 1/3n$

Solving after the symmetric indefinite factorization

multiplications:	$\text{NRHS}\,[n^2 + n]$
additions:	$\text{NRHS}\,[n^2 - n]$
total flops:	$\text{NRHS}\,[2n^2]$

Unblock QR factorization

multiplications:	$mn^2 - 1/3n^3 + mn + 1/2n^2 + 23/6n$
additions:	$mn^2 - 1/3n^3 + 1/2n^2 + 5/6n$
total flops:	$2mn^2 - 2/3n^3 + mn + n^2 + 14/3n$

Solving with the Unblock QR factorization

multiplications:	NRHS $[2mn - 1/2n^2 + 5/2n]$
additions:	NRHS $[2mn - 1/2n^2 + 1/2n]$
total flops:	NRHS $[4mn - n^2 + 3n]$

Bibliography

[1] J. O. Aasen. On the reduction of a symmetric matrix to tridiagonal form. *BIT*, 11:233–242, 1971.

[2] G. Alaghband. Parallel pivoting combined with parallel reduction and fill-in control. *Parallel Computing*, 11:201–221, 1989.

[3] F.L. Alvarado. Manipulation and visualization of sparse matrices. *ORSA J. Comput*, 2:186–207, 1989.

[4] G. Amdahl. The validity of the single processor approach to achieving large scale computing capabilities. In *Proceedings of the AFIPS Computing Conference*, volume 30, pages 483–485, 1967.

[5] P. Amestoy and I. S. Duff. Vectorization of a multiprocessor multifrontal code. *Int. J. Supercomputer Applic.*, 3:41–59, 1989.

[6] E. Anderson, Z. Bai, C. Bischof, J. Demmel, J. J. Dongarra, J. Du Croz, A. Greenbaum, S. Hammarling, and D. Sorensen. LAPACK working note #20: LAPACK: A portable linear algebra library for high-performance computers. Computer Science Technical Report CS-90-105, University of Tennessee, May 1990.

[7] E. Anderson and J. J. Dongarra. LAPACK working note #18: Implementation guide for LAPACK. Computer Science Technical Report CS-90-101, University of Tennessee, April 1990.

[8] P. Arbenz and G. Golub. On the spectral decomposition of Hermitian matrices modified by row rank perturbations with applications. *SIAM J. Matrix Anal. Appl.*, 9(1):40–58, January 1988.

[9] M. Arioli and I. S. Duff. Experiments in tearing large sparse systems. Technical Report CSS 217, Harwell Report, 1988. To appear in Proceedings of Advances in Numerical Computation, held at the National Propulsion Laboratory in July 1987 in memory of James Wilkinson.

[10] S. F. Ashby. CHEBYCODE: A Fortran implementation of Manteuffel's adaptive Chebyshev algorithm. Report UIUCDCS-R-85-1203, University of Illinois, Urbana, IL, May 1985.

[11] C. Ashcraft. A vector implementation of the multifrontal method for large sparse, symmetric positive definite linear systems. Report ETA-TR-51, ETA, 1987.

[12] O. Axelsson. Solution of linear systems of equations: Iterative methods. In V. A. Barker, editor, *Sparse Matrix Techniques: Copenhagen*, pages 1–51, Berlin, 1977. Springer-Verlag.

[13] O. Axelsson and G. Lindskog. On the eigenvalue distribution of a class of preconditioning methods. *Numer. Math.*, 48:479–498, 1986.

[14] I. Babuska. Numerical stability in problems of linear algebra. *SIAM J. Numer. Anal.*, 9(1):53–77, 1972.

[15] R. E. Benner, G. R. Montry, and G. G. Weigand. Concurrent multifrontal methods: shared memory, cache, and frontwidth issues. *Int. J. Super. Applic.*, 1:26–44, 1987.

[16] H. Berryman, J. Saltz, W. Gropp, and R. Mirchandaney. Krylov methods preconditioned with incompletely factored matrices on the CM-2. ICASE Report 89-54, NASA Langley Research Center, Hampton, VA, 1989.

[17] C. Bischof and C. Van Loan. The WY representation for products of Householder matrices. *SIAM J. Sci. Statist. Comput.*, 8(2):s2–s13, March 1987.

[18] A. Bjorck and T. Elfving. Accelerated projection methods for computing pseudo-inverse solutions of systems of linear equations. *BIT*, 19:145–163, 1979.

[19] J. Bunch and L. Kaufman. Some stable methods for calculating inertia and solving systems of linear equations. *Math. Comp.*, 31:163–179, 1977.

[20] D. Calahan, J. J. Dongarra, and D. Levine. Vectorizing compilers: A test suite and results. In *Supercomputer 88*, pages 98–105. IEEE Press, 1988.

[21] S. C. Chen, D. J. Kuck, and A. H. Sameh. Practical parallel band triangular system solvers. *ACM Trans. Math. Softw.*, 4:270–277, 1978.

[22] A. T. Chronopoulos and C. W. Gear. s-step iterative methods for symmetric linear systems. *J. Comput. Appl. Math.*, 25(2):153–168, 1989.

[23] P. Concus, G. H. Golub, and G. Meurant. Block preconditioning for the conjugate gradient method. *SIAM J. Sci. Statist. Comput.*, 6(1):220–252, 1985.

[24] A. K. Dave and I. S. Duff. Sparse matrix calculations on the CRAY-2. *Parallel Computing*, 5:55–64, 1987.

[25] T. A. Davis and P-C. Yew. A stable nondeterministic parallel algorithm for general unsymmetric sparse LU factorization. Report CSRD-908, University of Illinois, 1989. To appear in SIMAX.

[26] J. Demmel, J. J. Dongarra, J. Du Croz, A. Greenbaum, S. Hammarling, and D. Sorensen. Prospectus for the development of a linear algebra library for high-performance computers. Mathematics and Computer Science Division Report ANL-MCS-TM-97, Argonne National Laboratory, September 1987.

[27] D. Dodson and J. Lewis. Issues relating to extension of the Basic Linear Algebra Subprograms. *ACM SIGNUM Newsletter*, 20(1):2–18, 1985.

[28] J. J. Dongarra, editor. *Experimental Parallel Computing Architectures*. North-Holland, New York, 1987.

[29] J. J. Dongarra. Performance of various computers using standard linear equations software in a Fortran environment. Computer Science Technical Report CS-89-85, University of Tennessee, March 1990.

[30] J. J. Dongarra, J. Bunch, C. Moler, and G. Stewart. *LINPACK Users' Guide*. SIAM Pub., Philadelphia, 1979.

[31] J. J. Dongarra, J. DuCroz, I. Duff, and S. Hammarling. A set of Level 3 Basic Linear Algebra Subprograms. *ACM Trans. Math. Softw.*, 16:1–17, 1990.

[32] J. J. Dongarra, J. DuCroz, S. Hammarling, and R. Hanson. An extended set of Fortran Basic Linear Algebra Subprograms. *ACM Trans. Math. Softw.*, 14:1–17, 1988.

[33] J. J. Dongarra and I. S. Duff. Advanced architecture computers. Report CS-89-90, University of Tennessee, November 1989.

[34] J. J. Dongarra and S. C. Eisenstat. Squeezing the most out of an algorithm in Cray Fortran. *ACM Trans. Math. Softw.*, 10(3):221–230, 1984.

[35] J. J. Dongarra and E. Grosse. Distribution of mathematical software via electronic mail. *Communications of the ACM*, 30(5):403–407, July 1987.

[36] J. J. Dongarra, F. Gustavson, and A. Karp. Implementing linear algebra algorithms for dense matrices on a vector pipeline machine. *SIAM Rev.*, 26:91–112, January 1984.

[37] J. J. Dongarra and A. Hinds. Unrolling loops in Fortran. *Software—Practice and Experience*, 9:219–226, 1979.

[38] J. J. Dongarra and R. Hiromoto. A collection of parallel linear equation routines for the Denelcor HEP. *Parallel Computing*, 1:133–142, 1984.

[39] J. J. Dongarra and L. Johnsson. Solving banded systems on a parallel processor. *Parallel Computing*, 5:219–246, 1987.

[40] J. J. Dongarra, A. Karp, K. Kennedy, and D. Kuck. 1989 Gordon Bell Prize. *IEEE Software*, pages 100–110, May 1990.

[41] J. J. Dongarra and D. Sorensen. A portable environment for developing parallel Fortran programs. *Parallel Computing*, 5:175–186, July 1987.

[42] J. J. Dongarra, D. Sorensen, K. Connolly, and J. Patterson. Programming methodology and performance issues for advanced computer architectures. *Parallel Computing*, 8:41–58, 1988.

[43] J. J. Dongarra and D. C. Sorensen. Linear algebra on high-performance computers. In U. Schendel, editor, *Proceedings of Parallel Computing '85*, pages 3–32, New York, 1986. North Holland.

[44] J. J. Dongarra and D. C. Sorensen. A fully parallel algorithm for the symmetric eigenvalue problem. *SIAM J. Sci. Statist. Comput.*, 8(2):s139–s154, March 1987.

[45] P. Dubois, A. Greenbaum, and G. H. Rodrigue. Approximating the inverse of a matrix for use in iterative algorithms on vector processors. *Computing*, 22:257–268, 1979.

[46] P. Dubois and G. Rodrigue. An analysis of the recursive doubling algorithm. In Kuck et al., editors, *High speed computer and algorithm organization*, New York, 1977. Academic Press.

[47] I. S. Duff. MA28—A set of Fortran subroutines for sparse unsymmetric linear equations. Report R8730, HMSO, AERE Harwell, 1977.

[48] I. S. Duff. MA32—A package for solving sparse unsymmetric systems using the frontal method. Report R10079, HMSO, AERE Harwell, 1981.

[49] I. S. Duff. Direct methods for solving sparse systems of linear equations. *SIAM J. Sci. Statist. Comput.*, 5:605–619, 1984.

[50] I. S. Duff. The solution of sparse linear systems on the CRAY-1. In J. S. Kowalik, editor, *High-Speed Computation, NATO ASI Series*, volume F7, pages 293–309, Berlin, 1984. Springer-Verlag.

[51] I. S. Duff. Data structures, algorithms and software for sparse matrices. In D. J. Evans, editor, *Sparsity and Its Applications*, pages 1–29. Cambridge University Press, 1985.

[52] I. S. Duff. Parallel implementation of multifrontal schemes. *Parallel Computing*, 3:193–204, 1986.

[53] I. S. Duff. The parallel solution of sparse linear equations. In W. Handler, D. Haupt, R. Jeltsch, W. Juling, and O. Lange, editors, *CONPAR 86, Lecture Notes in Computer Science*, volume 237, pages 18–24, Berlin, 1986. Springer-Verlag.

[54] I. S. Duff. Multiprocessing a sparse matrix code on the Alliant FX/8. *J. Computational and Applied Math.*, 27:229–239, 1988.

[55] I. S. Duff, A. M. Erisman, C. W. Gear, and J. K. Reid. Some remarks on inverses of sparse matrices. Report CSS 171, AERE Harwell, 1985.

[56] I. S. Duff, A. M. Erisman, and J. K. Reid. *Direct methods for sparse matrices*. Oxford University Press, London, 1986.

[57] I. S. Duff, N. I. M. Gould, M. Lescrenier, and J. K. Reid. The multifrontal method in a parallel environment. Report CSS 211, Harwell, 1987.

[58] I. S. Duff, P. C. Jackson, A. Mills, and P. C. Robinson. Private communication.

[59] I. S. Duff and S. L. Johnsson. Node orderings and concurrency in structurally-symmetric sparse problems. In G. F. Carey, editor, *Parallel Supercomputing: Methods, Algorithms and Applications*, pages 177–189. Wiley, 1988.

[60] I. S. Duff and G. A. Meurant. The effect of ordering on preconditioned conjugate gradient. *BIT*, 29:635–657, 1989.

[61] I. S. Duff and J. K. Reid. MA27—A set of Fortran subroutines for solving sparse symmetric sets of linear equations. Technical Report R10079, HMSO, AERE Harwell, 1982.

[62] I. S. Duff and J. K. Reid. The multifrontal solution of indefinite sparse symmetric linear systems. *ACM Trans. Math. Softw.*, 9:302–325, 1983.

[63] I. S. Duff and J. K. Reid. The multifrontal solution of unsymmetric sets of linear systems. *SIAM J. Sci. Statist. Comput.*, 5:633–641, 1984.

[64] I. S. Duff, J. K. Reid, and J. A. Scott. The use of profile reduction algorithms with a frontal code. *Int. J. Num. Math. and Eng.*, 28:2555–2568, 1989.

[65] I.S. Duff, R.G. Grimes, and J.G. Lewis. Sparse matrix test problems. *ACM Trans. Math. Softw.*, 15:1–14, 1989.

[66] R. P. Dupont, T. Kendall and H. Rachford. An approximate factorization procedure for solving self-adjoint elliptic difference equations. *SIAM J. Numer. Anal.*, 53:559–573, 1968.

[67] S. C. Eisenstat. Efficient implementation of a class of preconditioned conjugate gradient methods. *SIAM J. Sci. Statist. Comput.*, 2:1–4, 1981.

[68] S. C. Eisenstat, H. C. Elman, and M. H. Schultz. Variational iterative methods for nonsymmetric systems of linear equations. *SIAM J. Numer. Anal.*, 20:345–357, 1983.

[69] S. C. Eisenstat, M. C. Gursky, M. H. Schultz, and A. H. Sherman. Yale sparse matrix package. 1: The symmetric codes. *Int. J. Num. Methods in Eng.*, 18:1145–1151, 1982.

[70] Engineering, Scientific Subroutine Library-Guide, and Reference. Release 3, Order No. sc23-0184, IBM, Kingston, NY, 1988.

[71] R. Fletcher. *Conjugate Gradient Methods for Indefinite Systems, Lecture Notes in Mathematics*, volume 506. Springer-Verlag, Berlin, 1976.

[72] M. Flynn. Very high speed computing systems. *Proc. IEEE*, 54:1901–1909, 1966.

[73] G. E. Forsythe and E. G. Strauss. On best conditioned matrices. *Proc. Amer. Math. Soc.*, 6:340–345, 1955.

[74] K. Gallivan, W. Jalby, U. Meier, and A. H. Sameh. Impact of hierarchical memory systems on linear algebra algorithmic design. *The Int. Journal of Supercomputer Appl.*, 21:12–48, 1988.

[75] K. Gallivan, R. Plemmons, and A. Sameh. Parallel algorithms for dense linear algebra computations. *SIAM Review*, 32(1):54–135, 1990.

[76] B. S. Garbow, J. M. Boyle, J. J. Dongarra, and C. B. Moler. *Matrix Eigensystem Routines— EISPACK Guide Extension, Lecture Notes in Computer Science*, volume 51. Springer-Verlag, New York, 1977.

[77] A. George, M. Heath, J. Liu, and E. Ng. Solution of sparse positive definite systems on a hypercube. Report ORNL/TM-10865, Oak Ridge National Laboratory, 1988.

[78] A. George, M. Heath, J. Liu, and E. Ng. Sparse Cholesky factorization on a local-memory multiprocessor. *SIAM J. Sci. Statist. Comput.*, 9:327–340, 1988.

[79] A. George and M. T. Heath. Solution of sparse linear least squares problems using Givens rotations. *Lin. Alg. & Appl.*, 34:69–83, 1980.

[80] A. George and J. W. H. Liu. *Computer Solution of Large Sparse Positive-Definite Systems.* Prentice-Hall, Englewood Cliffs, NJ, 1981.

[81] A. George and J. W. H. Liu. The evolution of the minimum degree ordering algorithm. *SIAM Review*, 31:1–19, 1989.

[82] A. George and E. Ng. SPARSPAK : Waterloo sparse matrix package user's guide for SPARSPAK-B. Department of Computer Science CS-84-37, University of Waterloo, 1984.

[83] A. George and E. Ng. An implementation of Gaussian elimination with partial pivoting for sparse systems. *SIAM J. Sci. Statist. Comput.*, 6:390–409, 1985.

[84] G. H. Golub and C. F. Van Loan. *Matrix Computations.* The Johns Hopkins Press, Baltimore, Maryland, 2nd edition, 1989.

[85] R. G. Grimes, D. R. Kincaid, and D. M. Young. ITPACK 2.0 user's guide. Technical report, Univ. of Texas, Austin, Texas, 1979.

[86] J. Gustafson. Reevaluating Amdahl's law. *Comm. ACM*, 31:532–533, 1988.

[87] J. Gustafson, G. R. Montry, and R. E. Benner. Development of parallel methods for a 1024-processor hypercube. *SIAM J. Sci. Statist. Comput.*, 9:609–638, 1988.

[88] I. Gustafsson. A class of first order factorization methods. *BIT*, 18:142–156, 1978.

[89] L. A. Hageman and D. M. Young. *Applied Iterative Methods.* Academic Press, New York, 1981.

[90] K. Hayami and N. Harada. The scaled conjugate gradient method on vector processors. In S. P. Kartashev and I. S. Kartashev, editors, *Supercomputing Systems, Proc. of the First International Conference, St.Petersburg*, pages 213–221, 1985.

[91] D. Heller. Some aspects of the cyclic reduction algorithm for block tridiagonal linear systems. *SIAM J. Numer. Anal.*, 13:484–496, 1978.

[92] J. Hennessy and D. Patterson. *Computer Architecture A Quantitative Approach.* Morgan Kaufmann Publishers, Inc., San Mateo, CA, 1990.

[93] M. R. Hestenes and E. Stiefel. Methods of conjugate gradients for solving linear systems. *J. Res. Natl. Bur. Stand.*, 49:409–436, 1954.

[94] R. Hockney and C. Jesshope. *Parallel Computers: Architecture, Programming and Algorithms.* Adam Hilger, Ltd., Bristol, United Kingdom, 1981.

[95] R. W. Hockney and I. J. Curington. $f_{1/2}$: A parameter to characterize memory and communication bottlenecks. *Parallel Computing*, 10:277–286, 1989.

[96] P. Hood. Frontal solution program for unsymmetric matrices. *Int. J. Num. Meth. Eng.*, 10:379–400, 1976.

[97] M. J. Hopper. Harwell subroutine library: A catalogue of subroutines, 9th ed. Report AERE R-9185, Harwell, 1989.

[98] Y. Huang and H. A. van der Vorst. Some observations on the convergence behaviour of GMRES. Report 89–09, Delft University of Technology, 1989.

[99] K. Hwang and F. Briggs. *Computer Architecture and Parallel Processing.* McGraw-Hill, New York, NY, 1984.

[100] B. M. Irons. A frontal solution program for finite-element analysis. *Int. J. Num. Meth. Eng.*, 2:5–32, 1970.

[101] K. C. Jea and D. M. Young. Generalized conjugate-gradient acceleration of nonsymmetrizable iterative methods. *Lin. Alg. & Appl.*, 34:159–194, 1980.

[102] E. Jessup and D. Sorensen. A parallel algorithm for computing the singular value decomposition of a matrix. Mathematics and Computer Science Division Report MCS-TM-102, Argonne National Laboratory, Argonne, IL, December 1987.

[103] O. G. Johnson, C. A. Micheli, and G. Paul. Polynomial preconditioning for conjugate gradient calculations. *SIAM J. Numer. Anal.*, 20:363–376, 1983.

[104] T. L. Jordan. A guide to parallel computation and some CRAY-1 experiences. Report LA-UR-81-247, Los Alamos National Laboratory, Los Alamos, NM, 1981.

[105] E. F. Kaasschieter. The solution of non-symmetric linear systems by bi-conjugate gradients or conjugate gradients squared. Report 86-21, Delft University of Technology, 1986.

[106] E. F. Kaasschieter. A practical termination criterion for the conjugate gradient method. *BIT*, 28:308–322, 1988.

[107] A. Karp and R. Babb. A comparison of 12 parallel Fortran dialects. *IEEE Software*, pages 52–67, 1988.

[108] L. Kaufman. Usage of the sparse matrix programs in the PORT library. Technical Report Report 105, Bell Laboratories, Murray Hill, New Jersey, 1982.

[109] R. Kettler. *Linear multigrid methods in numerical reservoir simulation.* PhD thesis, Delft University of Technology, Delft, 1987.

[110] D. R. Kincaid, T. C. Oppe, J. R. Respess, and D. M. Young. ITPACKV 2C User's Guide. Report CNA-191, Center for Numerical Analysis, Univ. of Texas, Austin, TX, 1984.

[111] P. Kogge. *The Architecture of Pipelined Computers.* McGraw-Hill, New York, 1981.

[112] A. E. Koniges and D. V. Anderson. ILUBCG2: A preconditioned biconjugate gradient routine for the solution of linear asymmetric matrix equations arising from 9-point discretizations. *Computer Physics Communications*, 43:297–302, 1987.

[113] J. J. Lambiotte and R. G. Voigt. The solution of tridiagonal linear systems on the CDC-STAR-100 computer. Technical report, ICASE-NASA Langley Research Center, Hampton, VA, 1974.

[114] C. Lawson, R. Hanson, D. Kincaid, and F. Krogh. Basic Linear Algebra Subprograms for Fortran usage. *ACM Trans. Math. Softw.*, 5:308–329, 1979.

[115] J. M. Levesque and J. W. Williamson. *A Guidebook to Fortran on Supercomputers.* Academic Press, 1989.

[116] J. W. H. Liu. The minimum degree ordering with constraints. *SIAM J. Sci. Statist. Comput.*, 10:1136–1145, 1988.

[117] J. W. H. Liu. The role of elimination trees in sparse factorization. *SIAM J. Matrix Anal. and Applics.*, 11:134–172, 1990.

[118] R. Lucas, T. Blank, and J. Tiemann. A parallel method for large sparse systems of equations. *IEEE Trans. on Computer-Aided Design*, CAD-6:981–991, 1987.

[119] N. K. Madsen, G. H. Rodrigue, and J. I. Karush. Matrix multiplication by diagonals on a vector/parallel processor. *Inform. Process. Lett.*, 52:41–45, 1976.

[120] T. A. Manteuffel. The Tchebychev iteration for nonsymmetric linear systems. *Numer. Math.*, 28:307–327, 1977.

[121] H. M. Markowitz. The elimination form of the inverse and its application to linear programming. *Management Sci.*, 3:255–269, 1957.

[122] K. K. Mathur and S. L. Johnsson. The finite element method on a data parallel computing system. *Int. J. of High-Speed Computing*, 1(1):29–44, May 1989.

[123] U. Meier. A parallel partition method for solving banded systems of linear equations. *Parallel Computing*, 2:33–43, 1985.

[124] U. Meier and A. Sameh. The behavior of conjugate gradient algorithms on a multivector processor with a hierarchical memory. Report 758, University of Illinois, Urbana, 1988.

[125] J. A. Meijerink. Iterative methods for the solution of linear equations based on incomplete factorisations of the matrix. Technical Report 643, Shell, KSEPL, Rijswijk, 1983.

[126] J. A. Meijerink and H. A. van der Vorst. An iterative solution method for linear systems of which the coefficient matrix is a symmetric M-matrix. *Math. Comp*, 31:148–162, 1977.

[127] J. A. Meijerink and H. A. van der Vorst. Guidelines for the usage of incomplete decompositions in solving sets of linear equations as they occur in practical problems. *J. Comp. Physics*, 44:134–155, 1981.

[128] G. Meurant. The block preconditioned conjugate gradient method on vector computers. *BIT*, 24:623–633, 1984.

[129] G. Meurant. Numerical experiments for the preconditioned conjugate gradient method on the CRAY X-MP/2. Report LBL-18023, University of California, Berkeley, 1984.

[130] G. Meurant. The conjugate gradient method on vector and parallel supercomputers. Report CTAC-89, University of Brisbane, 1989.

[131] C. Moler. A closer look at Amdahl's law. Technical Report TN-02-687, Intel, 1987.

[132] J. Ortega and C. Romine. The *ijk* forms of factorization II. Parallel systems. *Parallel Computing*, 7(2):149–162, 1988.

[133] C. C. Paige and M. A. Saunders. LSQR: An algorithm for sparse linear equations and sparse least squares. *ACM Trans. Math. Softw.*, 8:43–71, 1982.

[134] G. Radicati and Y. Robert. Vector and parallel CG-like algorithms for sparse non-symmetric systems. Report 681-M, IMAG/TIM3, Grenoble, France, 1987.

[135] G. Radicati and M. Vitaletti. Sparse matrix-vector product and storage representations on the IBM 3090 with Vector Facility. Report G513-4098, IBM-ECSEC, Rome, 1986.

[136] J. K. Reid. Sparse matrices. In D. A. H. Jacobs, editor, *The State of the Art in Numerical Analysis*, pages 85–146, New York, 1977. Academic Press.

[137] J. R. Rice and R. F. Boisvert. *Solving Elliptic Problems Using ELLPACK*. Springer-Verlag, Heidelberg, 1985.

[138] Y. Saad. Practical use of polynomial preconditionings for the conjugate gradient method. *SIAM J. Sci. Statist. Comput.*, 6:865–881, 1985.

[139] Y. Saad. Krylov subspace methods on supercomputers. Report September 19, RIACS, Moffett Field, 1988.

[140] Y. Saad and M. H. Schultz. GMRES: A generalized minimal residual algorithm for solving nonsymmetric linear systems. *SIAM J. Sci. Statist. Comput.*, 7:856–869, 1986.

[141] J. J. F. M. Schlichting and H. A. van der Vorst. Solving bidiagonal systems of linear equations on the CDC CYBER 205. Report NM-R8725, CWI, Amsterdam, 1987.

[142] J. J. F. M. Schlichting and H. A. van der Vorst. Solving 3D block bidiagonal linear systems on vector computers. *Journal of Comp. and Appl. Math*, 27:323–330, 1989.

[143] R. Schreiber and C. Van Loan. A storage efficient WY representation for products of Householder transformations. *SIAM J. Sci Stat Comp.*, 10(1):53–57, 1989.

[144] M. K. Seager. Parallelizing conjugate gradient for the CRAY X-MP. *Parallel Computing*, 3:35–47, 1986.

[145] A. H. Sherman. Algorithm 533. NSPIV, A Fortran subroutine for sparse Gaussian elimination with partial pivoting. *ACM Trans. Math. Softw.*, 4:391–398, 1978.

[146] H. D. Simon, P. Vu, and C. Yang. Performance of a supernodal general sparse solver on the CRAY Y-MP: 1.68 Gflops with autotasking. Report SCA-TR-117, Boeing Computer Services, Seattle, 1989.

[147] S. W. Sloan and M. F. Randolph. Automatic element reordering for finite-element analysis with frontal schemes. *Int. J. Num. Meth. Eng.*, 19:1153–1181, 1983.

[148] B. T. Smith, J. M. Boyle, J. J. Dongarra, B. S. Garbow, Y. Ikebe, V. Klema, and C. Moler. *Matrix Eigensystem Routines—EISPACK Guide., 2nd ed.*, volume 6. Springer-Verlag, New York, 1976.

[149] P. Sonneveld. CGS, A fast Lanczos-type solver for nonsymmetric linear systems. *SIAM J. Sci. Statist. Comput.*, 10:36–52, 1989.

[150] D. Sorensen and C. Van Loan. Block factorizations for symmetric indefinite matrices, Cornell technical report. In preparation.

[151] G. W. Stewart. *Introduction to Matrix Computations*. Academic Press, New York, 1973.

[152] H. Stone. *High Performance Computer Architecture*. Addison-Wesley, New York, 1987.

[153] H. L. Stone. Iterative solution of implicit approximations of multidimensional partial differential equations. *SIAM J. Numer. Anal.*, 5:530–558, 1968.

[154] H. S. Stone. An efficient parallel algorithm for the solution of a tridiagonal linear system of equations. *JACM*, 20:27–38, 1973.

[155] W. F. Tinney and J. W. Walker. Direct solutions of sparse network equations by optimally ordered triangular factorization. *IEEE Proc.*, 55:1801–1809, 1967.

[156] A. van der Sluis. Condition numbers and equilibration of matrices. *Numer. Math.*, 14 1:14–23, 1969.

[157] A. van der Sluis and H. A. van der Vorst. The rate of convergence of conjugate gradients. *Numer. Math.*, 48:543–560, 1986.

[158] A. van der Sluis and H. A. van der Vorst. Numerical solution of large sparse linear algebraic systems arising from tomographic problems. In G. Nolet, editor, *Seismic Tomography*, Dordrecht, 1987. Reidel.

[159] H. A. van der Vorst. A vectorizable variant of some ICCG methods. *SIAM J. Sci. Statist. Comput.*, 3:350–356, 1982.

[160] H. A. van der Vorst. The performance of Fortran implementations for preconditioned conjugate gradients on vector computers. *Parallel Computing*, 3:49–58, 1986.

[161] H. A. van der Vorst. Large tridiagonal and block tridiagonal linear systems on vector and parallel computers. *Parallel Computing*, 5:45–54, 1987.

[162] H. A. van der Vorst. The convergence behavior of some iterative solution methods. In R. Gruber, J. Periaux, and R. P. Shaw, editors, *Proc. of the Fifth Int. Symp. on Numer. Methods in Engng.*, volume 1, 1989.

[163] H. A. van der Vorst. High performance preconditioning. *SIAM J. Sci. Statist. Comput.*, 10:1174–1185, 1989.

[164] H. A. van der Vorst. ICCG and related methods for 3D problems on vector computers. *Computer Physics Communications*, 53:223–235, 1989.

[165] H. A. van der Vorst. Experiences with parallel vector computers for sparse linear systems. *Supercomputer*, 37:28–35, 1990.

[166] H. A. van der Vorst and K. Dekker. Vectorization of linear recurrence relations. *SIAM J. Sci. Statist. Comput.*, 10:27–35, 1989.

[167] R. S. Varga. *Matrix Iterative Analysis.* Prentice-Hall, Englewood Cliffs, NJ, 1962.

[168] P. K. W. Vinsome. ORTOMIN: An iterative method for solving sparse sets of simultaneous linear equations. In *Proc. Fourth Symposium on Reservoir, Simulation, Society of Petroleum Engineers of AIME*, pages 149–159, 1976.

[169] H. F. Walker. Implementation of the GMRES method using Householder transformations. *SIAM J. Sci. Statist. Comput.*, 9:152–163, 1988.

[170] H. H. Wang. A parallel method for tridiagonal equations. *ACM Trans. Math. Softw.*, pages 170–183, 1981.

[171] W. Ware. The ultimate computer. *IEEE Spectrum*, pages 89–91, March 1973.

[172] D. T. Winter. Efficient use of memory and input/output. In J. J. te Riele, T. J. Dekker, and H. A. van der Vorst, editors, *Algorithms and Applications on Vector and Parallel Computers*, Amsterdam, 1987. North-Holland.

[173] Z. Zlatev, J. Wasniewski, and K. Schaumburg. *Y12M - Solution of Large and Sparse Systems of Linear Algebraic Equations*, volume 121. Springer-Verlag, New York, 1981.

Index